BROOKLANDS BOOKS

LAND ROVER
SERIES ONE
1948-1958

Compiled by
R.M. Clarke

ISBN 1 85520 0805

 Brooklands Books Ltd.
'Holmerise', Seven Hills Road,
Cobham, Surrey, England
Printed in Hong Kong

BROOKLANDS BOOKS

CONTENTS

ACKNOWLEDGEMENTS

This is the fourth in our series of Brooklands Books devoted to Land Rovers, and with it we complete the set — at least, for the moment. Our other volumes cover the Series II, Series III, and 90/110 models.

Material on the Series I Land-Rover is hard to come by, and we are grateful to John Smith, Tony Hutchings, David Bowyer, Dave Shephard and James Taylor for the loan from their collections of some of the rare material we have reproduced here. James Taylor has also kindly penned a few words of introduction for us, and Dave Shephard supplied our cover picture.

Like all Brooklands Books, this one could not have been put together without the generosity and understanding of those who hold the copyright to the articles it reproduces. Our thanks on this occasion go to the management of The Ambassador, Autocar, Classic and Sportscar, Commercial Motor, Farmers Weekly, The Field, Land Rover Ltd, Land Rover Owner, Motor, Motor Trader, Motor Transport, Motor Trend, Old Motor, Off Road & 4 Wheel Drive, Practical Classics, Thoroughbred & Classic Cars and The Woodworker.

R.M. Clarke

The Land-Rover owed its conception totally to the after-effects of the Second World War. In the late 1940s, the Rover Company had to have a vehicle to sell abroad if it was to be granted enough Government-rationed steel to survive as a motor manufacturer. At the same time, it saw a new market opening up, as farmers who had encountered motor vehicles perhaps for the first time when conscripted, began to see their potential in agriculture. Both at home and abroad, demobbed wartime Jeeps were proving popular for use on the farm, and Rover decided to exploit this market by building their own Jeep which, they hoped, would attract the necessary overseas sales.

The Rover "jeep" was designed to use as many of the company's existing car components as possible and to be panelled with aluminium alloy, which had proven its worth in aircraft during the War. Its construction was to be simple, to save the cost of tooling, and it was to be both rugged and cheap. By April 1948, the first pilot-production examples were ready, and Rover displayed them at the Amsterdam Motor Show.

Rover's intention was that its new Land-Rover should be a stop-gap product, built only until normality returned to the car market. But the new vehicle proved so popular that, by 1951, twice as many Land-Rovers as cars were coming off the Rover production lines. In fact, nearly one and a half million vehicles later, the Land-Rover is still with us.

The articles in this book provide a fascinating insight into the first ten years of one of Britain's proudest exports. In that period, the vehicle gradually evolved to meet the demands of its market, by offering more powerful engines, more economical engines, and larger payloads. Yet at the end of those ten years, the Land-Rover was still recognisably the same vehicle which had been drawn up so hurriedly during 1947 — a tribute to the soundness of its original design.

I am sure Land-Rover enthusiasts everywhere will join me in welcoming this latest Brooklands title, which is certain to become an essential source-book on the marque.

James Taylor

THE LAND-ROVER

Entirely New Multi-purpose Vehicle Designed to Offer Go-anywhere Transport, a Portable Source of Power, and an Alternative to the Light Tractor

WAR-TIME experience with light four-wheel-drive vehicles gave a plain indication of the wide scope that exists in all parts of the world for a go-anywhere vehicle with a plain utility-type body which can be used both for personal or goods transport.

Inevitably, such vehicles have a particular appeal to agriculturists, whether in this country or abroad, and their needs have received particular attention in an outstandingly interesting vehicle of this kind, announced this week by the Rover Co., Ltd. Known as the Land-Rover, it combines the go-anywhere properties already mentioned with many of the qualities of a light tractor, plus the added scope offered by a portable source of power which is available either for operating plant actually mounted on the vehicle or for driving external farm or industrial machinery.

Leading features of the Land-Rover include a 1,595 c.c. four-cylinder petrol engine (as used in the new Rover "60" car), a transmission system which incorporates a transfer box giving both two or four-wheel drive and alternative sets of gear ratios, provision for a power take-off both at the rear and in the centre of the chassis, and provision also for a winch at the front if desired. The body is of the utility-type built up of non-corrodible light-metal panelling, but sturdily reinforced with steel at all points liable to bear the brunt of rough usage.

Robust Frame

The chassis frame of the Land-Rover is a particularly massive structure of the all-welded type with side members and five main cross-members fully boxed. All are of heavy-gauge steel, and the robust construction is indicated by the fact that the main side members are 6 ins. deep at the centre and 3 ins. wide throughout. Another noteworthy point is that the cross-members at the extreme ends are carried out to the full width, both to serve the purpose of bumpers and to provide a particularly rigid structure for towing or for the mounting of any auxiliary equipment.

Gears and Drive

The four-cylinder engine is of the overhead inlet, side exhaust type, as in the Rover "60" car, with the split between the cylinder head and block inclined, an unusual head shape giving excellent breathing and enabling a high compression ratio to be used on low-grade fuel without pinking. In the Land-Rover the compression ratio has been lowered very slightly but, in other respects, the engine is as described in the issue of "The Motor" dated February 18, 1948. It is rubber mounted at four points in the chassis and offset approximately 2¼ ins.

In unit with the engine is a four-speed synchromesh gearbox, the internal components of which (apart from a slightly closer first gear ratio) are identical with those of the car. To the rear of this, however, is bolted a further casing which forms part of the transfer box. The primary purpose of this is to convey the drive to the offset propeller shafts which lead forward and aft to the front and rear axles. It also serves the important functions of giving an alternative set of gear ratios and providing arrangements for disconnecting the front-wheel drive when this is not required. Also incorporated is a free wheel, the purpose of which will be referred to later.

The drive is conveyed from the tailshaft of the gearbox via a pair of intermediate pinions to the final pair of pinions mounted on the shaft connected to the two propeller-shafts. Of the latter pinions, one, which provides the normal upper set of ratios, is freely mounted on its shaft, whilst the other is slidably mounted on splines; in its forward position it engages with internal dogs and locks the upper ratio pinion to its shaft, so bringing the higher set of ratios into action. In its rearmost position it provides an overall reduction

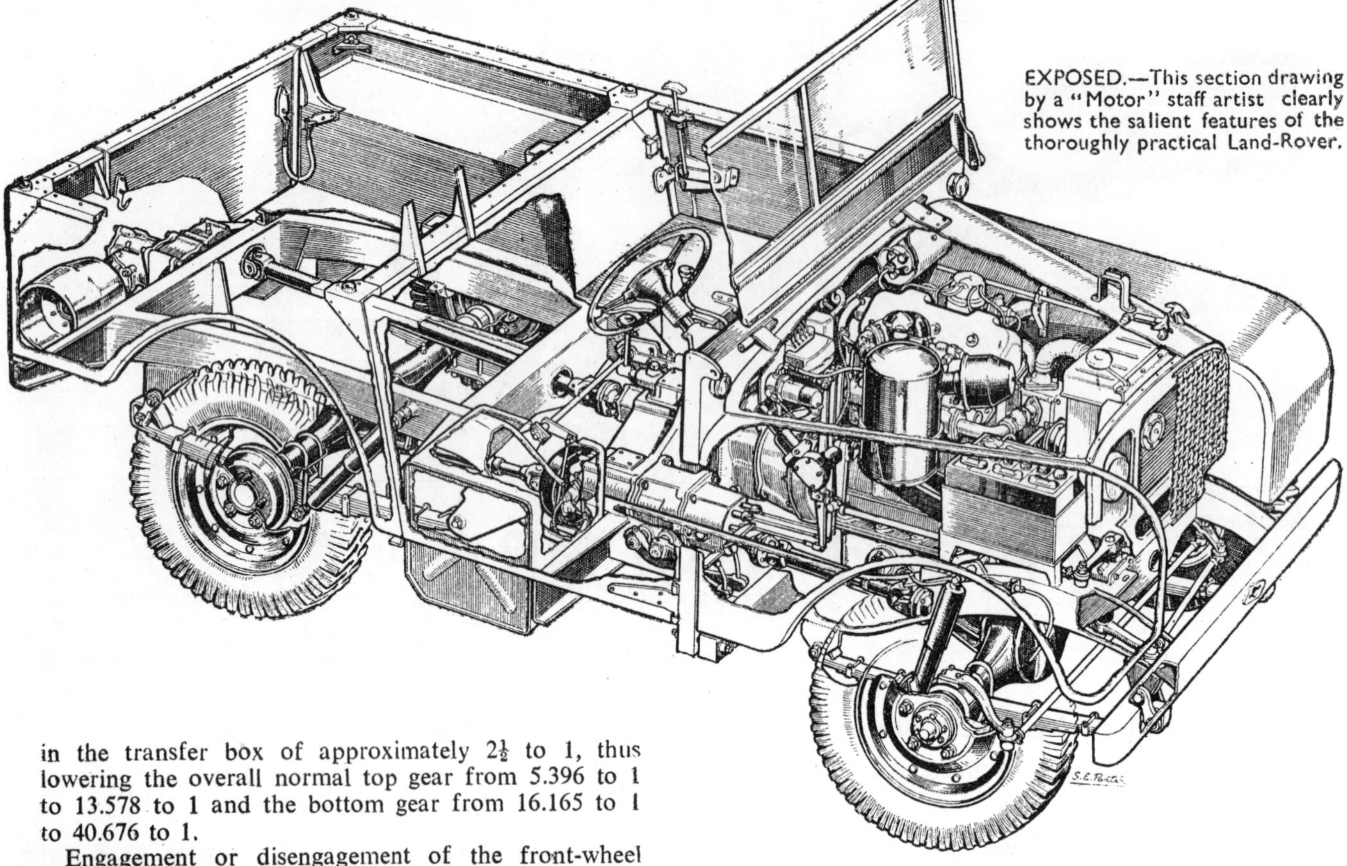

EXPOSED.—This section drawing by a "Motor" staff artist clearly shows the salient features of the thoroughly practical Land-Rover.

in the transfer box of approximately 2½ to 1, thus lowering the overall normal top gear from 5.396 to 1 to 13.578 to 1 and the bottom gear from 16.165 to 1 to 40.676 to 1.

Engagement or disengagement of the front-wheel drive is effected by a dog clutch and also incorporated is the free wheel already mentioned. The purpose of this is to allow the front wheels to over-run the rear, a very necessary function since any deviation of the vehicle from the straight results in the front wheels following a path of larger radius than the rear. On soft or loose ground this would be of comparatively little importance because it could readily be accommodated by a small degree of wheel slip. This, however, does not apply when the vehicle is driven on firm road surfaces, on which a high skid torque would be built up within the transmission; this would not only put an undesirable strain on the mechanism (and make disengagement of the drive almost impossible owing to the " winding-up " effect), but would also finally expend itself in tyre slip to the detriment of tread life.

The incorporation of a free wheel in a four-wheel-drive vehicle is a very notable refinement, because avoidance of the undesirable effects just mentioned otherwise depends entirely on the driver remembering to disconnect the front drive before reaching firm surfaces. The free wheel can be locked to provide four-wheel drive in reverse.

Brakes and Springs

Mounted on an extension of the final shaft of the transfer box is a transmission brake of the Girling type operating in a 9-in. drum. Owing to the action of the free wheel it takes effect only on the rear wheels when travelling forward. The foot-brake operates on all four wheels through a conventional Girling hydraulic system.

Both axles incorporate conventional spiral bevel drive to live half-shafts with, in the case of the front wheels, constant speed universals.

Torque reaction is taken through the springs, which are semi-elliptic all round, controlled by telescopic hydraulic shock absorbers, an interesting point being that the second leaf of each spring is extended at each end to surround (with a suitable clearance) the spring eye of the master leaf, thus acting as a safety factor in the unlikely event of a main-leaf breakage. In all cases rubber bushes are embodied, both for the shock-absorber mountings and for the spring eyes, thus eliminating all

COVERED WAGON. — Various forms of weather-protection are available for the Land-Rover. This one has the metal doors, side windows and full canvas hood fitted, the latter having a rear curtain with a celluloid window.

6

lubrication points in the suspension system.

Reduction of maintenance is also in evidence in the steering system, in which all ball joints are packed with lubricant on assembly and sealed, so that no attention whatever is required until the time finally comes for a general overhaul. A detail point in connection with the sealing is that the rubber caps employed are retained by coiled spring circlips, so that should the rubber stretch and harden in use the seal will be unaffected. The steering layout is designed so that a right- or left-hand driving position can be arranged with equal ease.

Controls are conventional except for certain additions brought about by the auxiliary features of the vehicle. The main gear lever works after the conventional fashion, and the normal clutch, brake and accelerator pedals are provided. Additional controls take the form of three push-pull knobs, which protrude from the dash below and to the right of the instrument panel. These control the high and low ratio in the transfer box, the free-wheel lock and the front-drive engagement dog. All are normally operated when the vehicle is station-

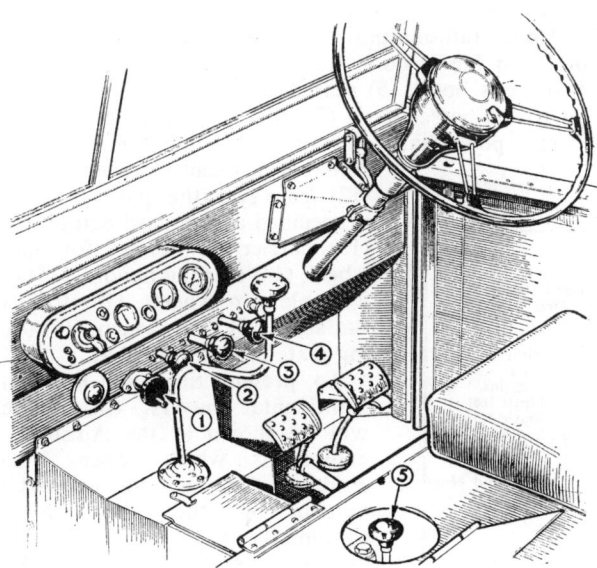

KNOBS AND PEDALS.—This sketch shows the controls for: 1. Hand throttle; 2. Transfer gear; 3. Freewheel; 4. Front wheel drive; 5. Power take-off

ary. An additional lever (which is situated below a hinged flap in the seat ramp to the side of the driver's seat) controls the rear power take-off.

Instruments include a speedometer, petrol gauge and ammeter, whilst warning lights are provided for the engine lubrication system, the choke (the light shines so long as this is in action) and the ignition. A useful detail is provision for plugging in an inspection lamp,

and there is also a hand-throttle control designed to operate over the entire power range for use when the engine is providing stationary power

Power take-off arrangements, as already indicated, are three in number To the rear of the main gearbox approximating roughly to the position of a normal transmission propeller shaft, is a shaft leading to a power take-off mounted on the rear cross-member, external to the vehicle This can be supplied either with a splined shaft for a direct power take-off or with a pulley to provide a flat belt drive. In the latter case, a pair of helical gears, conveys the drive from the main shaft to a short shaft terminating in a spiral bevel, which drives a corresponding bevel on the pulley shaft

For the provision of external power this arrangement fulfils most purposes, but where the Land-Rover is required for operating portable plant actually mounted on the vehicle a centre power take-off is available, the drive in this case being taken from behind the gearbox.

In addition, the Land-Rover can be obtained with a capstan-type winch mounted on the front cross-member. In this case the drive is taken from the nose of the crankshaft via a worm and wheel, giving a right-angle drive for the vertical capstan and providing a reduction in the neighbourhood of 30 to 1 Engagement is by means of dogs

Practical Layout

Bodywork on the Land-Rover is severely practical. The general arrangement is clear from the accompanying illustrations and need not be described in detail. Dimensions, however, are interesting The internal width measured between the cappings of the doors and body sides is $56\frac{1}{2}$ ins., and the width of the floor at the rear is $34\frac{1}{2}$ ins. between the raised portions above the wheels. The latter rise 9 ins from the floor, and the distance between their upper surfaces and the top of the body sides is a further $5\frac{1}{2}$ ins

In a fore-and-aft direction the length of the floor is $42\frac{1}{4}$ ins., but a portion of this is taken up by the spare wheel, which sits in a well just behind the division between the front and rear compartments When desired, however, the spare wheel can be transferred to a horizontal position on the bonnet top. which is specially strengthened for the purpose, thus leaving the rear entirely unobstructed The unladen height of the floor from the ground is 20 ins at the front and 27 ins at the rear

The doors, which taper slightly towards the base, have a maximum width of 34 ins., and interesting details are the hinges (which are of the gate type, allowing the doors to be readily detached if desired or folded back flush with the straight-sided front wings) and the latches (which are operated by pull-up handles).

Except for the steel bulkhead behind the engine, the screen frame and the various fittings, the body is entirely constructed of heavy-gauge aluminium alloy to eliminate corrosion, but the body sides, doors, partition between front and rear compartments and the let-down tailboard are all surmounted by heavy steel cappings to withstand hard use. These cappings, like all the other external steel fittings, are galvanized to prevent rust.

The windscreen is extremely large, rising to a total height of 23¼ ins. above the bonnet, the two glass panels each measuring 27 ins. wide and 15 ins. deep. When required, the screen can be folded flat over the bonnet, where it rests on supports provided and is held in place by spring catches.

Weather protection is available in the form of either a driver's hood or a hood for the complete vehicle, the latter having plain fabric sides and an openable rear curtain incorporating a celluloid window. For the front doors, rigid side screens are available or, alternatively, the purchaser can specify combined fabric doors and side screen on a metal frame, a safety strap being provided in this case.

Extras to Order

In a vehicle of this kind, which is specifically designed for a wide variety of purposes, it is evident that different

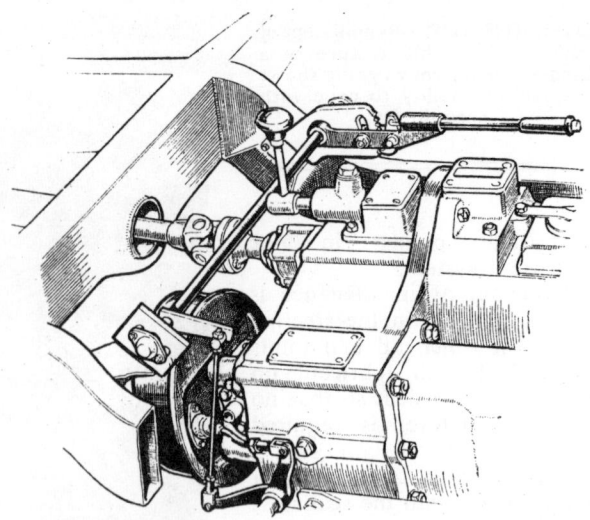

GIRLING GRIP.—Mounted on an extension of the final shaft of the transfer box, this Girling hand-operated transmission brake takes effect on the rear wheels only, when travelling forward.

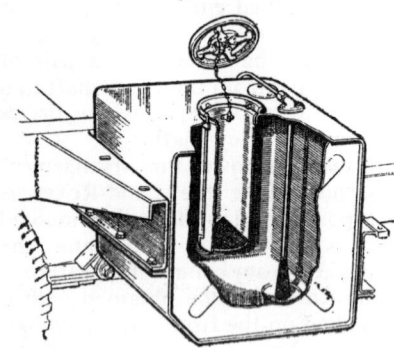

Details of the fuel filters are shown in this sketch of the Land-Rover's ten-gallon fuel tank, situated under the driver's seat.

purchasers will require different items of equipment, according to the use to which the machine is to be put. Accordingly, the Rover Co. has adopted the wise policy of quoting a basic price for the bare vehicle, this figure including such essentials as full range of instruments, lighting equipment, windscreen and driver's seat, but nothing which is likely to be unwanted by some buyers. The price of the vehicle in this state is £450, and the buyer is thus in the happy position of being required to pay only for those additional items of equipment which are necessary for his particular needs.

Prices of additional equipment are not yet available, but a list of the main items offered is as follows:—(1) Doors (aluminium), (2) side screens for above, (3) combined fabric door and side screen on metal frame, (4) driver's hood, (5) rear hood, (6) cushion and backrest (front passengers), (7) heater, (8) engine governor, (9) rear power take-off, (10) pulley unit for rear power take-off, (11) centre power take-off, (12) spare tyre (the spare wheel only is specified as standard equipment), (13) carrier on bonnet for spare wheel, (14) starting handle, (15) detachable rim wheels, (16) tropical radiator, (17) front winch, (18) towing plate for rear drawbar.

To present the Land-Rover to Continental buyers, two models will be shown at the Amsterdam exhibition which opens next Friday. One will be a standard version but the other will be equipped with an arc-welding unit driven by vee-belts from the centre power take-off.

In launching this new vehicle the Rover Co. has displayed an enterprise which should be well rewarded, and there is no doubt that a big market, both at home and abroad, exists for a machine such as the Land-Rover. There is no doubt, also, that in its design the Rover Co. has applied a wide knowledge and experience not only of vehicle manufacture, but of agricultural and industrial requirements

ROVER "LAND-ROVER"

Engine Dimensions :		Transmission —contd.	
Cylinders	4	Prop. shafts	Hardy Spicer
Bore	69.5 mm.	Final drive	Spiral bevel
Stroke	105 mm.		(4-wheel drive)
Cubic capacity	1,595 c.c.	**Chassis Details :**	
Piston area	23.5 sq. ins.	Brakes	Girling
Valves	O.H. Inlet, Side Exhaust	Brake drum diameter	10 ins.
Compression ratio	6.8 to 1	Friction lining area	94.25 sq. ins.
		Suspension, front	Semi-elliptic leaf
Engine Performance :		Suspension, rear	Semi-elliptic leaf
Max. b.h.p.	50/55	Shock absorbers	Telescopic hydraulic type
at	4,000 r.p.m.	Wheel type	Split Rim
Max. b.m.e.p.	125	Tyre size	6.00 x 16
at	2,000 r.p.m.	Steering gear	Burman, Worm and Nut
B.H.P. per sq. in.		Steering wheel	Spring Spoke
piston area	2.13/2.34		
Peak piston speed ft.		**Dimensions :**	
per min.	2,760	Wheelbase	6 ft. 8 ins.
		Track, front	4 ft. 2 ins.
Engine Details :		Track, rear	4 ft. 2 ins.
Carburetter	Solex	Overall length	11 ft. 0 in.
Ignition	Coil	Overall width	5 ft. 0½ in.
Plugs : make and type	Lodge, 14 mm.	Overall height	6 ft. 0 in.
Fuel Pump	S.U. Electric	Ground clearance	8¾ ins.
Fuel capacity	10 gallons	Turning circle	33 ft. 0 in.
Oil filter	By-pass	Dry weight	2,398 lb.
Oil capacity	10 pints		
Cooling system	Pump and fan	**Performance Data :**	
Water capacity	19 pints	Piston area, sq. ins.	
Electrical system	12 volts	per ton	22.0
Battery capacity	51 amp.-hours	Brake lining area, sq.	
		ins. per ton	88
Transmission :		Top gear m.p.h. per	
Clutch	Borg and Beck, OR	1,000 r.p.m.	High, 15.5 ; Low, 6.2
	Newton and Bennet	Top gear m.p.h. at	
Gear ratios : Top	High, 5.396 ; Low, 13.578	2,500 ft. / min.	
3rd	High, 8.039 ; Low, 20.229	piston speed	High, 56.4 ; Low, 22.4
2nd	High, 11.023 ; Low, 27.738	Litres per ton-mile,	
1st	High, 16.165 ; Low, 40.676	dry	High, 23.00 ; Low, 58.00
Rev.	High, 13.743 ; Low, 34.587		

15 CWT. BROCKHOUSE TRAILER FOR
The LAND-ROVER

OVERRUN AND PARKING BRAKE
REVERSING STOP
PATENTED BALL COUPLING
ADJUSTABLE DROP LEG
COLD ROLLED STEEL CHANNEL CHASSIS
PRESSED STEEL BODY
DOMED WINGS
WHEELS AND TYRES INTERCHANGEABLE WITH LAND-ROVER

FOR GREATLY INCREASED CARRYING CAPACITY

This 15 cwt. Brockhouse Trailer designed for use with the Land-Rover will be found of inestimable use in many ways. It immediately makes available a greatly increased carrying capacity. It is strongly constructed of steel but light in weight and easily and quickly attached and detached. Incorporated in the design is a combined over-run and parking brake. For the Farmer, the Market Gardener and for general use, this combination of the Land-Rover and Brockhouse Trailer offers a carrying service of wide utility.

SPECIFICATION

CAPACITY : 15 cwts.
BODY : Sides and ends 18 gauge Pressed Steel Panels. Floor 16 gauge.
BODY DIMENSIONS : 6'×3' 2"×1' 6" sides.
SPRINGS : 2" wide Silico Manganese Steel. Anchored at front and shackled at rear. Rolled eyes Greaser type shackle and spring eye bolts.
AXLE : 1½" sq. bed axle in good quality mild steel.
WHEEL HUBS : Drop-forged steel fitted with ample size roller bearings.

WHEELS AND TYRES : Pressed steel disc wheels interchangeable with Land-Rover wheels.
HITCH : Brockhouse patent ball-hitch made from drop-forged steel components, and incorporating over-run brake mechanism.
BRAKES : 8" dia. cam operated two shoe brakes mounted on pressed steel back plate, operated by overrun mechanism. A patented parking brake is provided on the drawbar.
PAINTING : Chassis and body, green to match Land-Rover.
TAIL LIGHT : Rear light & Stop light complete with flex. Number plate & "T" plate provided. **PRICE £75 · 0 · 0**

OTHER **LAND-ROVER** LITERATURE *you may care to write for . . .*

1. The "LAND-ROVER STATION WAGON" the ideal all-purpose vehicle—which meets all occasions.
2. The "LAND-ROVER" for the Farmer, Countryman and General Industrial use.

THE ROVER COMPANY LIMITED
SOLIHULL · BIRMINGHAM · ENGLAND

Telephone : Sheldon 2461. Telegrams : Rover, Solihull.

Service Depot : SOLIHULL. Telegrams : Rovrepair, Solihull. London Showrooms : DEVONSHIRE HOUSE, PICCADILLY, W.1. Telephone : Grosvenor 2092. London Service Depot : SEAGRAVE ROAD, FULHAM, S.W.6. Telephone : Fulham 1221. Telegrams : Rovrepair, Phone Fulham.

● E. & O.E. : Specifications and price subject to alteration without notice.

The name Land-Rover is a registered Trade mark of The Rover Company Limited.

The LAND-ROVER
WITH ALL-WEATHER EQUIPMENT

THE standard production Land-Rover with its all-weather equipment erected comprising extra strong serviceable hood with rear panel, laced for easy detachment. Two aluminium side doors with sliding Perspex sidescreens.

● No allowance can be made for any item of standard equipment not required. Specification and prices subject to alteration without notice.

GO anywhere... DO anything

● **ENGINE.** High efficiency four-cylinder Capacity 1595 c.c. Develops more than 50 B.H.P. 25-27 m.p.g.

● **CHASSIS.** Side and cross members of box section. Light but exceptionally rigid.

● **BODYWORK.** High tensile non-corrodible aluminium sheet metal work.

● **POWER TAKE-OFF.** Gives a powerful pulley drive for generators, compressors or agricultural equipment.

● **ELECTRICAL SYSTEM.** 12-volt starting and lighting.

The versatility of the Land-Rover is really amazing. A four-wheel drive tractor, a delivery wagon, a mobile power plant and a fast economical vehicle on the road — the Land-Rover is all these things rolled into one. With its power take-off, that can be coupled up to any equipment needing pulley drive, it makes a direct appeal to farmers, field engineers, industrialists, in fact anyone who needs a fast, powerful, adaptable, utility vehicle. The Land-Rover is built for hard work and hard wear at low running costs and (note for the exporter) is supplied with right or left-hand drive as required. Ask your local dealer for particulars.

Price of vehicle only without additional equipment — £450.

LAND-ROVER
Britain's most versatile vehicle

Made by The Rover Company Limited, Solihull, Birmingham.

9

A Maid-of-all-work for the Farmer

IN the Land-Rover, which has just been announced, The Rover Co.. Ltd., of Solihull, Birmingham, has combined the functions of a light van or a car and trailer, with those of a light tractor, thus providing a single vehicle which will meet the majority of needs of the small farmer, or provide for the stand-by requirements of owners of larger farms. The vehicle also offers wide scope for works transport, and as a portable source of power.

These widely varying requirements are catered for by a sturdily built machine, powered by a four-cylindered petrol engine, linked to a transmission system providing two- or four-wheel drive (alternative sets of gear ratios at will), a choice of power take-off arrangements, and a front winch. Mounted on this multi-purpose chassis is a utility-type body in which non-corrodible light-metal panelling is freely used, with stout steel reinforcements at all points liable to rough treatment.

The power unit, which is of 1,595 c.c. capacity (69.5 mm. × 105 mm.), has an unusual arrangement of overhead inlet and side-exhaust valves, and is, in fact, similar to that used in the recently introduced Rover " 60 " private car. Unusual features of the engine design comprise an inclined split between block and head, which makes possible a valve layout embodying an exhaust valve in the cylinder block at

This picture gives an excellent impression of the general sturdiness of the frame construction. The shaft running to the rear power take-off will be noted.

(Above) The Land-Rover revels in negotiating terrain such as here shown. It can be driven at high speed on cross-country work. (Below) A variety of equipment is offered with the vehicle, a complete hood being amongst the extras obtainable.

55 degrees from the vertical, and an inlet valve in the head at 20 degrees from the vertical. Operation is from a single camshaft, the exhaust valves being actuated through the medium of short rockers, and the overhead inlet valves by conventional push rods and rockers.

This arrangement makes for good breathing and valve cooling, excellent anti-pinking qualities and mechanical stiffness. The power output is 50-55 b.h.p. at 4,000 r.p.m.

With the needs of stationary work in view, special attention has been paid to adequate cooling, the fan being cowled to ensure a good cooling draught through the radiator block. The latter is exceptionally well protected in that it is not only situated well back from the front of the vehicle, but is also protected by a substantial metal pressing, and by a stout meshed grille; the headlamps also share this protection.

In unit construction with the engine,

is a conventional dry-plate clutch, a four-speed synchromesh gearbox and the transfer box which is on the off-side of the main gearbox. This complete assembly is off-set in the frame by approximately 2¼ ins. to provide space for the transfer box, and for the drive shafts leading fore and aft to the front and rear axles respectively.

The drive to the transfer box is taken via a pair of intermediate pinions which mesh with a further pair of gearwheels on the transfer-box mainshaft. Of the two wheels on the latter, one (which

gives the high set of ratios) is mounted freely on its shaft, and is provided with external dogs corresponding to the splines on which the second wheel is slidably mounted.

Thus, by moving the sliding pinion in one direction the high-rate pinion becomes locked to the transfer-box mainshaft, bringing the upper set of ratios into operation. Moving the pinion in the other direction brings its teeth into mesh with the appropriate intermediate pinion, and provides the low set of ratios.

At the rear, the transfer-box main-shaft is coupled direct to the rear propeller shaft but, in the case of the front drive, a dog clutch is interposed to provide for disconnection. In addition, there is an important refinement in the shape of a free-wheel. Normally, it is desirable that the drive to the front wheels be disconnected when a vehicle of this type is used on firm road surfaces, but the introduction

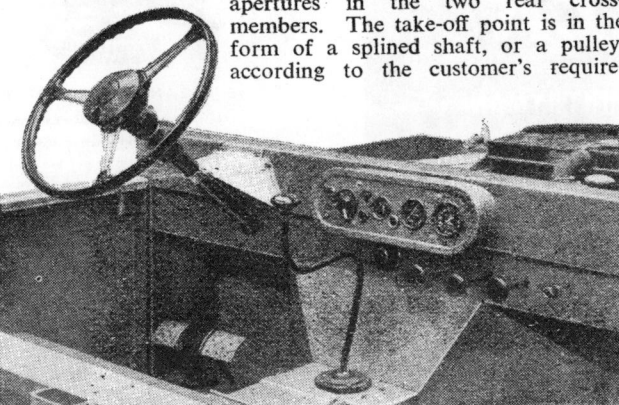

The driver's controls and instrument panel on the export version of the Land-Rover. In this essentially practical vehicle embellishment takes second place to ease and convenience of handling.

of a free-wheel removes the possibility of serious consequences following neglect of this precaution.

A conventional spiral-bevel final drive, incorporating a differential, is embodied in each axle and, in the case of the front drive, constant-speed universals are employed to provide for steering movement. Driving torque is taken through the semi-elliptic springs which are not only provided with rubber bushes to reduce maintenance, but a further interesting point is that the secondary leaf in each case is continued with a suitable clearance around the spring eye to prevent serious dislocation of the axles in the unlikely event of a main-leaf breaking.

Reduction of maintenance work is also to be noted in connection with the steering system, all the ball joints of which, on assembly, are packed with sufficient lubricant to last until a general overhaul becomes due. Sealing is effected by moulded rubber caps which fit into recesses, where they are located by circlips. The Burman worm-and-nut type steering box is mounted on the engine bulk-head, and right or left-hand drive is obtainable.

Box-section Frame

In order to provide for the high stresses to which the chassis frame of a vehicle of this type will inevitably be subjected, all the frame members are exceptionally robust. Heavy-gauge box-section steel is used for the side-members, which are 6 ins. deep in the centre of the frame, and 3 ins. wide throughout, and for the five main cross-members. Of the latter, those at the extreme front and rear are well extended, bumper fashion, to serve the dual purpose of protecting the vehicle, and providing substantial mounting points for towing brackets and auxiliaries. A Girling hydraulic braking system takes effect on all four wheels, and the hand brake works inside a 9-in. drum located behind the transfer box.

The design is such that power is available at three auxiliary points. At the front, a de-ditching winch can be fitted. This is driven from the nose of the crankshaft via a worm drive giving a reduction of approximately 30 to 1; it is engaged by means of dogs.

A power take-off can be mounted aft of the back cross-member, the drive being obtained from a shaft leading from the tail of the gearbox through apertures in the two rear cross-members. The take-off point is in the form of a splined shaft, or a pulley, according to the customer's require-

(Right) The power unit, which has several unusual features, develops 50-55 b.h.p. at 4,000 r.p.m. Side. exhaust valves and overhead inlet valves are used.

(Left) How the power take-off is arranged at the rear end of the Land-Rover. The drive is obtained through a shaft running direct from the tail-end of the gearbox.

ments. In addition, a second power take-off, in the centre of the chassis, is offered for purposes involving the use of portable plant carried on the vehicle itself.

The essentially practical type of body provides an internal width between door cappings of 4 ft. 8½ ins., whilst the rear compartment measures 3 ft. 6¼ ins. from front to rear, with a floor width between the raised portions above the wheels of 2 ft. 10½ ins. Floor height, with the vehicle unladen, is 1 ft. 8 ins. at the front, and 2 ft. 3 ins. at the rear.

Normally, the spare wheel is accommodated in the rear compartment where it occupies a vertical position

behind the driver's seat, but where the whole of the rear compartment is required for luggage, an alternative mounting above the reinforced bonnet top is offered.

The wide doors are mounted on gate-type hinges which permit them to be folded back flush against the front wings or, if desired, quickly removed; side screens are available if required.

As already indicated, corrosion has virtually been eliminated by the use of light aluminium alloy for the body, but the tops of the body sides and doors are stoutly reinforced with steel cappings, galvanized, as are all external ferrous fittings, to prevent rust. The windscreen is arranged to fold flat over the bonnet when not required. The price of the bare vehicle is £450.

Prices are not at present available for additional equipment, but a list of the main items which will be scheduled as extras is as follows: doors (aluminium) and sidescreens; combined fabric door and sidescreen on metal frame; driver's hood; rear hood; cushion and back-rest (front passenger); heater; engine governor; rear power-take-off and pulley unit; centre power-take-off; spare tyre (only the spare wheel is specified as standard equipment); carrier on bonnet for spare wheel; starting handle; detachable-rim wheels; tropical radiator; front winch, towing plate for rear drawbar.

Leading dimensions of the complete vehicle are as follow:—Wheelbase, 6 ft. 8 ins., track, 4 ft. 2 ins.; overall length, 11 ft., overall width, 5 ft. 0½ in.; overall height, 6 ft.; ground clearance, 8¾ ins., turning circle, 33 ft., dry weight, 2,398 lb

The LAND-ROVER

FOR THE FARMER, THE COUNTRYMAN AND GENERAL INDUSTRIAL USE

THE CLEAN DESIGN OF THE LAND-ROVER CHASSIS

Under chassis view of the Land-Rover showing the drive to the front and rear axles and the rear power take-off. The Land-Rover is not an adapted vehicle, it is specially designed and built to cover a wide field of usefulness in the service of agriculture and general industry. The exceptionally robust construction of the Land-Rover chassis is evident in the lower illustration. The depth of the box section chassis and the powerful engine will be noted. The drive to the rear power take-off is through the main gearbox. The front and rear axles are driven from the transfer box.

Here the Land-Rover is seen without the rear power take-off in use on the farm pulling an 8 ft. tandem-disc harrow.

Britain's Most LAND-ROVER

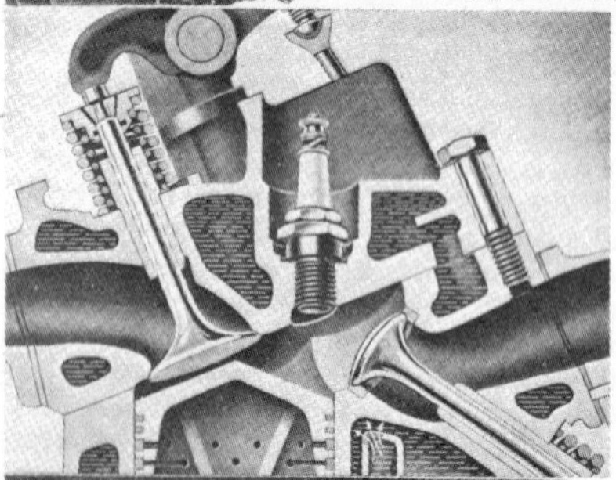

THE SECRET of the performance of the new Rover engine lies in this special cylinder head design. It is completely new and an example of advanced thought in automobile engineering. Test results show a remarkable improvement both in power and economy.

For taking produce to the Station or Market the Land-Rover is without equal. It is fast, economical and requires a minimum of attention.

A MOBILE POWER PLANT . . . . POWER JUST WHERE YOU WANT IT THROUGH THE POWER TAKE OFF

ACTION!

This is the Land-Rover as a fast economical vehicle. It is supplied with right or left hand steering as required

THE GO ANYWHERE VEHICLE

These pictures tell their own story

Fording streams or over exceptionally rough ground the Land-Rover goes everywhere. It is designed for those jobs that no ordinary vehicle could do. Four-wheel drive gives maximum possible traction and a transfer gearbox in combination with the main gearbox provides a range of eight forward speeds. A free-wheel incorporated with the front wheel propeller shaft acts as a differential to take care of varying ground conditions.

Designed to meet the urgent need for increased production in agriculture and industry, the Land-Rover is a truly multi-purpose vehicle. It is fast and economical for use on the road, and at the same time has that ability to keep going under the very worst cross-country conditions which can only be achieved by four-wheel drive.

It can be used as a light tractor, and with power take-off as a mobile power plant.

On the farm it will do much of the work for which a tractor is used. With its four-wheel drive it will transport heavy loads over ploughed fields or other places where the going is hard. As a mobile power unit it takes power to the job wherever it is. Through the rear power take-off it can be harnessed to drive the threshing machine, the elevator for rick building, the chaff cutter, the circular saw and countless other jobs that call for portable power. A centre power take-off can be provided for driving an air compressor for tree spraying or paint spraying and for portable milking apparatus.

It can be used to drive mobile welding plant, and in industry its uses range from moving heavy machinery and delivering goods to providing power in an emergency, and trucking around the factory. It has been designed with the same skill and experience and is built in the same factory as the world-famous Rover car.

THE ROVER COMPANY LIMITED
SOLIHULL, BIRMINGHAM, ENGLAND

Service Depot . . SOLIHULL   Telephone: Sheldon 2461   Telegrams: Rover, Solihull
Telegrams . . Rovrepair, Solihull   London Showrooms: London Service Depot: SEAGRAVE ROAD, FULHAM, S.W.6
     DEVONSHIRE HOUSE, PICCADILLY, W.1   Telephone Grosvenor 2092   Telephone   Fulham 1221
Reprinted by   Telegrams: Rovrepair, Phone, Fulham
The Land Rover Series One Club

Here the Land-Rover is seen driving a circular saw through the rear power take-off. The powerful transmission brake keeps the vehicle rock steady with no danger of movement.

Versatile Vehicle

SPECIFICATION

ENGINE. Flexibly mounted on rubber at four points. Four cylinders, bore 69·5 mm. Stroke 105 mm. 1595 c.c. Maximum brake horse power 50. Three bearing counterbalanced crankshaft of high specification steel of ample dimensions. Camshaft in crankcase driven by double roller silent chains with hydraulic adjuster. Firing order 1, 3, 4, 2.

VALVES. Overhead inlet valves operated by rocker and push rod from camshaft. Side exhaust valves with inserted valve seat operated by rocker in direct contact with camshaft.

PISTONS. Aluminium. Inverted "V" shaped head to conform to patented design of hemi-spherical combustion chamber giving increased compression turbulence. Two compression and two scraper rings are fitted.

CLUTCH. Single dry plate 9" diameter.

DYNAMO. Automatic voltage regulator 12 V.

STARTER. Operates on flywheel.

CARBURETTOR. Down-draught.

PETROL FILTER. Mounted on dash.

AIR CLEANER. Oil bath type.

TRANSMISSION. To rear and front axle by open propeller shaft via two speed transfer box.

REAR AXLE. Semi-floating. Spiral bevel type. Ratio 4·88:1.

COOLING SYSTEM. Water circulation by pump. Thermostatic control. A fan is fitted. Water capacity two gallons.

LUBRICATION. By pressure from gear type pump forcing oil to all bearings, timing chain and valve gear. Capacity 10 pints.

GEARS. Four forward speeds and reverse. Ratios: first 3·00:1, second 2·04:1, third 1·47:1, top 1:1, reverse 2·54:1.

TRANSFER BOX. Incorporates two speeds which in conjunction with the main gearbox give a comprehensive range of eight forward gears. Ratios: first 2·888:1, top 1·146:1.

IGNITION. Coil and battery. Automatic controlled ignition advance 12 volt battery. Capacity 52 amp. hours.

FRONT AXLE. Fitted with differential similar to rear axle. The drive to front wheels is through free-wheel and constant velocity universal joints totally enclosed.

CHASSIS. Side and cross members of box section forming an exceptionally rigid assembly.

STEERING. The steering wheel optional right or left hand driving position.

FUEL SUPPLY. From 10 gallon tank under driver's seat.

SPRINGS. Semi-elliptic. Four tubular type shock absorbers are fitted.

WHEELS. Detachable disc wheels having 4½" wide rims. Tyres 16 × 6·00 Heavy Duty traction type.

DIMENSIONS. Overall width 5' approx. Overall length 11' approx. Weight of vehicle 22¼ cwt. Wheelbase 80". Track 50".

DRAW BAR PULL. 1,200 lbs. to 1,800 lbs.

MAXIMUM ROAD SPEED. Over 50 m.p.h.

REAR POWER TAKE-OFF (at extra cost). Drive through back of main gearbox to rear of chassis. Can be fitted to give puley drive for threshers, chaff cutters, circular saw, etc., or shaft drive for power mowers, binders, combine harvesters, etc.

CENTRE POWER TAKE-OFF (at extra cost). Arranged to drive by V belts, compressors, generators and other portable equipment which can be mounted in the body.

BODY and general sheet metal work of high tensile non-corrodible light alloy.

ALL external steel fittings galvanised.

E. & O.E. Subject to alteration without notice.

Inset shows the Land-Rover in use as an inter-factory vehicle for transporting lines of trucks. It can be operated in confined spaces and is invaluable as a time-saving aid to production.

INDUSTRIAL OR FARM TRANSPORT

In its application for farm or industrial transport the Land-Rover possesses many advantages over the more commonly used vehicles. It will pull a heavy load over a ploughed field and give fast and economical operation on normal roads. It will go where most other vehicles cannot reach, and its running costs are low. The non-corrodible light alloy bodywork together with galvanised exterior steel-work ensures low maintenance charges and no deterioration of finish under the worst weather conditions.

SOME FEATURES of the
LAND-ROVER

High efficiency engine. develops over 50 B.H.P.

* * *

Box-section chassis frame, light, exceptionally rigid.

* * *

High tensile non-corrodible aluminium sheet metal work.

* * *

All exterior steel fittings galvanised.

* * *

Free-wheel to front axle acts as self-locking differential.

* * *

Special protection for ignition system against damp.

* * *

Powerful hydraulic brakes.

* * *

Wide front seat accommodates 3.

* * *

12-volt lighting and starting.

...left is the powerful high four-cylinder engine of the ...ver. Simple in design and of ...nstruction. It develops more ...h.p.

A power-driven winch of the capstan type can be mounted at the front of the Land-Rover. The winch can be applied to countless jobs from moving heavy machines in the factory to grubbing out old tree roots on the farm.

On the farm the Land-Rover is coupled through the rear power take-off to the elevator for rick building.

An All-purpose ROVER

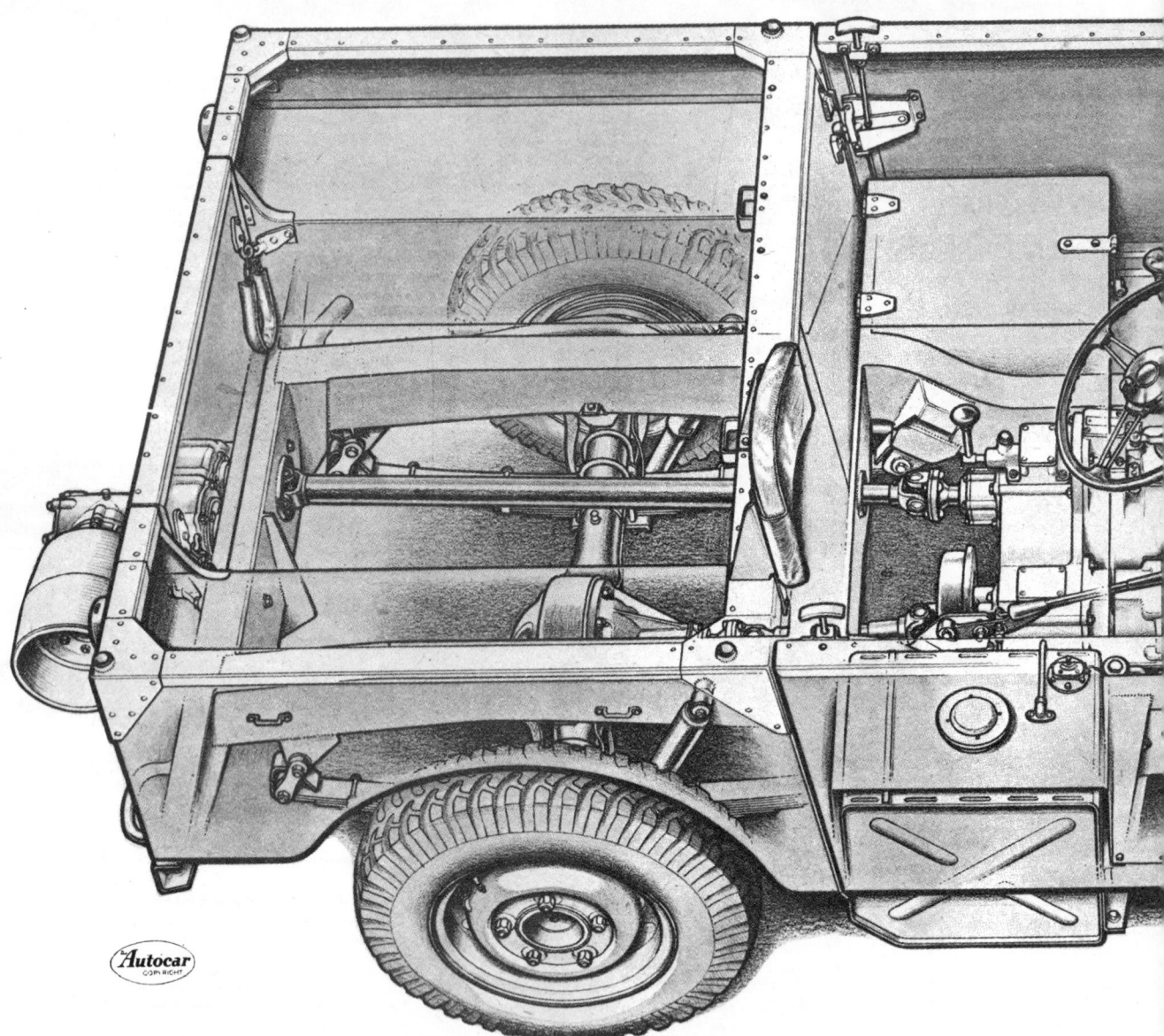

Autocar
COPYRIGHT

ALTHOUGH these pages are normally devoted to the many aspects of the purely private car and its usage, there is now something to describe which can either be regarded as a private car able to perform many most valuable duties other than sheer transport, or as a general purpose countryside worker which is also capable of providing comfortable and efficient transport. This dual role of the new Land-Rover, regardless of which range of duty is of the greater value to the owner, cannot be too highly stressed, because it opens up possibilities of the greatest value to

those who live in the country, whether under cultivation or in the wild state. So much has been said and written in the past about the so-called People's Car, much of it nonsense, that the advent of a really practical British vehicle which goes far beyond that over-publicized proposal should be hailed with genuine acclamation.

Let us consider the new Land-Rover in a sharp perspective, and see what it offers. It is a car with a wheelbase of 6ft 8in and a track of 4ft 2in; the overall length is 11ft, and the width 5ft. It has an open body able to seat three in a row on the front seat and

four, wagonette-style, in the back—seven in all. It can have doors, side panels, and a complete all-weather equipment. It has a modern design high-efficiency four-cylinder engine developing 50 b.h.p. On main roads it can average as high as 40 miles within each hour, and its fuel consumption at that fast average is round about 27 m.p.g. It weighs 22½ cwt. It has been designed and built by the Rover Company, and that in itself is a guarantee of quality which will be instantly accepted by any British motorist.

Its appearance is starkly practical;

Practical Road and Cross-country Vehicle Built to High Standards

This cutaway drawing by *The Autocar* artist John Ferguson shows in detail all the purposeful features of a strictly functional machine. Chief points are : Optional four-wheel drive via two-speed transfer box, with free wheel in the drive to the front ; new Rover o.h. inlet, side exhaust-valve engine ; rigid frame ; high ground clearance ; rear power take-off, and real accessibility. The appearance has an astringent freshness after an overdose of what have been unkindly described as " tinware balloons."

SPECIFICATION

Engine.—Rover Model 60. 4 cylinders, 69.5×105 mm (1,595 c.c.). Overhead inlet and side exhaust valves. V-shaped piston crowns giving increased compression turbulence. 50 b.h.p. Three-bearing counter-balanced crankshaft. Thermostat controlled water pump circulation. Downdraught carburettor; oil bath filter.

Transmission.—Dry single-plate clutch. 4-speed gear box with synchromesh on third and top. Four wheel drive. Power unit flexibly mounted. Overall gear ratios: Transfer box in high ratio, 5.6, 8.36, 11.46 and 16.8 to 1. Transfer box in low ratio, 14.12, 21.04, 28.85 and 42.3 to 1.

Electrical Equipment.—Lucas. Coil and battery ignition. 12-volt with constant voltage control dynamo. 52 ampere-hour battery. Inbuilt head lamps.

Tank Capacity.—10 gallons. Electric pump.

Suspension.—Half-elliptic springs with telescopic hydraulic dampers.

Brakes.—Girling hydraulic four-wheel, with transmission brake.

Wheels and Tyres.—Steel disc wheels with 4½in rims. Tyres 16×6.00in heavy-duty traction type.

Main Dimensions.—Wheelbase, 6ft 8in. Track, 4ft 2in. Overall length, 11ft; width, 5ft. Weight, 22½ cwt. Draw bar pull, 1,200 to 1,800 lb.

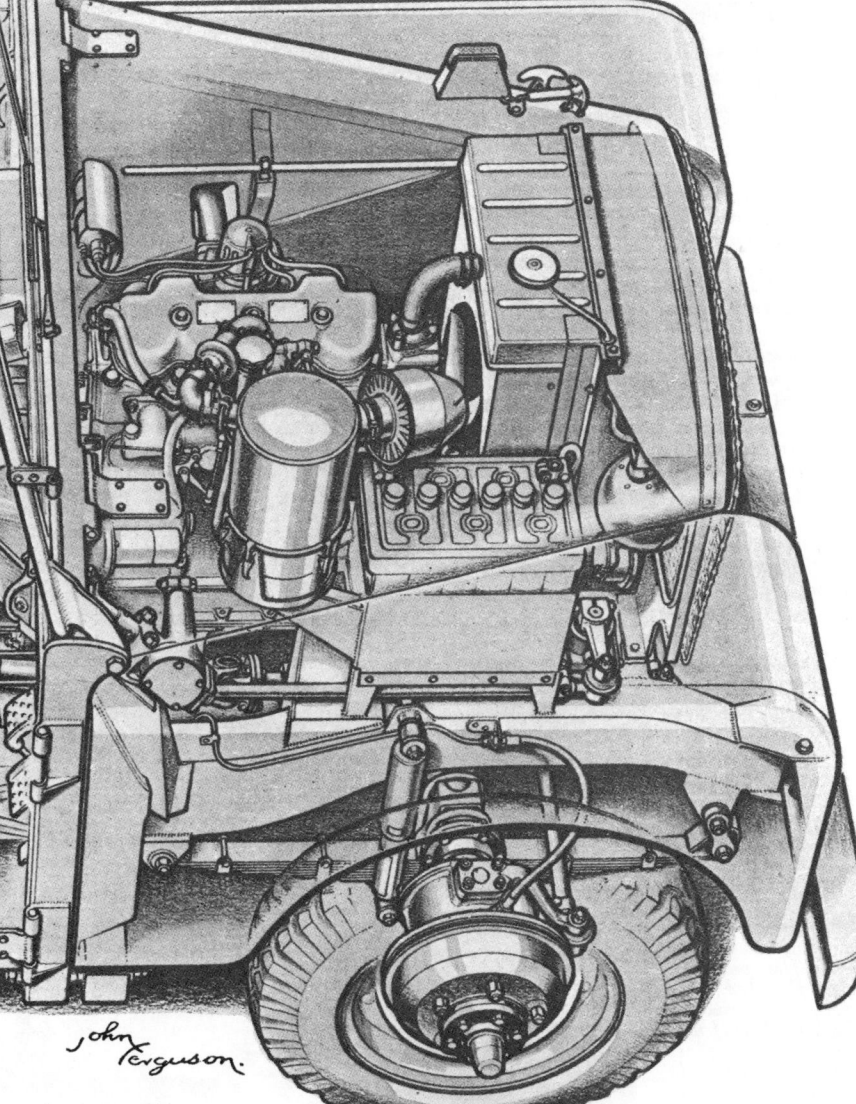

parts, front, middle, and rear, and any one of these parts can be removed in just about 15 minutes. The cylinder bores are chromium plated to prevent wear or corrosion, and the engine is fitted with a large size oil bath type of air filter, to keep out fine dust and road sand. Even the ignition wiring is waterproofed. The Land-Rover can go through floods up to its wheel centres or deeper without trouble. The operative word about the whole car is "substantial."

All these features alone make one think. If the world has to be strictly economical for years to come, is not this the sort of car that most of us need, one that is entirely practical and essentially usable? Washing is reduced to a minimum, and maintenance is easy; there is no carrying about of weight more or less uselessly devoted to fashionable appearance and not really essential luxury. And it is a car usable

Goods may be stowed " all over the place " in the Land-Rover when open—in the back, alongside the driver, and, no doubt, on the bonnet when there are " just a couple of sacks left over."

there is nothing of the luxury vehicle about its looks. Nevertheless it is not ugly and has a distinctly attractive appearance all its own. Everything necessary for travel is there. Accessibility reaches a degree which does not exist in an orthodox modern car, and the finish inside and out is of a kind which permits the owner to leave the vehicle in the open in any weather without fear of deterioration. All the exposed steel parts, including the chassis frame, are heavily galvanized with zinc, and the body panels and flooring are of a non-corroding hard aluminium alloy. The body is in three

Weather protection is complete when the driving compartment is closed up, and the deep screen and ample headroom will appeal to drivers who want to see where they are going. Head lamps and radiator are fully screened. Side lamps are scuttle-mounted.

in open form or completely enclosed. These points alone proclaim themselves.

But they tell only half the story. The Land-Rover is a mobile power station as well, and will tow or do a variety of useful work on the land over rough ground. It can drive a large circular saw and cut up timber for firewood. It can be used with trailers to transport heavy loads over ploughed fields or other hard going. As a mobile power unit it takes the power to the job. Through the power take-off it can be harnessed to drive a threshing machine, an elevator, or a chaff-cutter. It can draw a plough, and most other farm implements. It can perform all these tasks because it is specifically provided with a four-wheel drive, and a power take-off.

Drive Transmission

The engine, clutch, and gear box are of the same type as in the latest Rover 60 and 75 cars, but behind the tail of the gear box and to one side of it is a second transfer gear box. From the front of this a propeller-shaft runs to the front axle, which has a spiral bevel drive and differential gear, and from the back a second propeller-shaft runs to the rear axle, which has a drive similar to the front. A third shaft runs from the tail of the main box to the extreme tail of the car, where there can be mounted a power take-off gear box complete with a wide-belt pulley. The normal gear box provides four speeds and reverse, whilst the transfer box offers a choice of two different ratios, with a separate control, so the car virtually has eight gear ratios.

A rather interesting point is that the drive to the front wheels passes through a free wheel, for use on main roads, but locked by a separate control when four-wheel drive is needed in reverse. The rear power take-off is an extra, and another extra can be provided in the form of a centre power take-off for driving compressors, generators, or other portable equip-

ment which can be mounted in the body. Provision is also made for a power-driven winch of the capstan type mounted at the front of the vehicle, which can be used for a variety of jobs from moving machinery in a factory to grubbing out old tree roots on the farm. It is a vehicle of almost unlimited uses.

The Land-Rover has an immensely strong frame, of deep box section throughout, including the cross-members. The extreme ends are used to support stout bumper bars. Suspension back and front is by half-elliptic springs controlled by telescopic hydraulic dampers. Rubber bushes needing no lubrication are used throughout. The steering can be on either the left or the right side. The engine has the latest Rover overhead inlet

and side exhaust valves. Pressure water cooling is used, and air is circulated by an extra large enclosed fan. At an atmospheric temperature of 80 deg the engine can run at full output more or less indefinitely, and at a slightly lower output at 100 deg.

Perhaps the most surprising thing about this remarkable car is the price, £450 in its basic state; that is, without extra equipment. Purchase tax does not apply. It is hoped to reach a production figure of 200 vehicles a week by the end of this year. Normally the car has no doors, but these are available and can be dropped into special hinges instantly; they are also immediately removable. Above the doors there can be detachable metal frame side panels with sliding Perspex windows. These also drop into substantial sockets and can be firmly locked. The tubular frame for the hood is detachable—it does not fold, but is bodily removable. The covering can be left over the top, and the side curtains rolled up into it. The tail board also opens and is detachable. The screen can be folded flat forwards out of the way.

A spare wheel is carried inside the body behind the front seat, but if the full capacity of the body is needed the wheel can be moved on to an additional mounting provided on the top of the bonnet. Incidentally, when one lifts the bonnet one is gratified to see a finely finished engine with no suggestion of agricultural machinery about it.

The Land-Rover is likely to increase the name of quality British goods all over the world.

News of the Week

British manufacturers are taking an interest in jeeps. This week the Rover Co., Ltd., announced the production of a new vehicle of this type

Here is a new jeep which has been designed by Nuffield Mechanizations and is being tested by fighting-vehicle experts of the Ministry of Supply at Chobham. It has torsion-bar suspension, integral chassis and body construction, and a flat four-cylindered 60 b.h.p. engine. A power take-off for operating a winch, and a drawbar attachment are provided. The 24-volt electrical equipment includes a two-speed dynamo for a wireless set. The vehicle is said to be capable of a speed of 60 m.p.h.

E. H. Row and Joseph Lowrey go

LLAND ROVING

An Expedition into Wales with Rover's New All-purpose Vehicle

CLIMBING UP.—On the first section of the Monks' Way, where the old road clambers up out of the Elan Valley.

TWM SION CATTI was a Welshman. He lived in a cave near Llandovery and was, more or less, the Robin Hood of those parts. If he and his cave hadn't crept into the conversation the Ordnance maps would not have been brought out and, if they hadn't, Row and Lowrey wouldn't have spotted the dotted line straggling across the mountains between Rhayder and Ffair Rhos marked " Ancient Road."

Now, as readers of " The Motor " of rather more thán six months' standing may have realized, Messrs. Row and Lowrey have a weakness for ancient roads— the more ancient the road, the greater the weakness— and this one seemed to have all the makings. There was one section where the water-splashes came thick and fast and several points on the map where the contours, even at one inch to the mile, clustered so closely that you could barely get a pin between them. Additionally, should this way prove so ancient as to be completely decayed, there were plenty of other dotted lines in the neighbourhood which seemed to offer possibilities of some decidedly interesting " off-the-road " motoring. And, of course, apart from all this, there was Twm Sion Catti's cave, where, it was said by those who knew, one might even see a buzzard.

To attempt this sort of trip in an orthodox car was, quite obviously, out of the question unless (a) there was plent of time to spare, and (b) plenty of surplus energy ditto. Accordingly, overtures were made to the Rover Co. with a view to borrowing one of the new Land Rovers, which has drive to all four wheels and a transfer box providing a useful set of low ratios. The sugges-

CHARGING THROUGH.—Preliminary reconnaissance plus an abundance of enterprise saw the Land Rover through the earlier marshy spots.

tion appealed to them but, just to make sure that there should be no misapprehension about the capabilities of the car, we were given a demonstration run over their local demonstration circuit. Any misapprehensions were effectively removed. As a precaution, however, it was decided that a length of rope and a driving pulley for the power take-off be included for bollarding the car out if we did get it stuck.

Llandovery was chosen as the focal point for the expedition because, not only was it convenient to the selected area, but it was not far from the venue for a subsidiary project, namely, an assault upon a hill near Brechfa which enjoys the name of " Dunlop's Dividend " and had, according to the records, never been climbed clean by a four-wheeled, self-propelled vehicle.

Convenient Arrangement

A lovely June evening was happening as, in company with photographer Ross, we wended our way through the traffic-packed streets of Birmingham, three-abreast on individual seats, and westwards out on to the open road at a cruising speed around the 50-mark with the speedometer flicking up occasionally to over 60.

On trips of this sort there is a lot to be said for this three-abreast seating, because the navigator can stay put and give directions to the driver easily, while the photographer-cum-gate-opener can nip in and out smartly whenever occasion demands.

Early dinner at the " Royal Oak " at Leominster was a most satisfying affair, and by 9.30 p.m. the Land Rover was tucked away in its garage at Llandovery and we were enjoying our beer inside the Castle Hotel— our chosen rendezvous. And not such a bad choice, either, with, amongst other things, adequate and up-to-date plumbing altogether above the general standard of small, country-town hotels.

The following morning, there being all the signs of a beautiful day in prospect, the Rover was completely stripped of its all-weather equipment excepting the

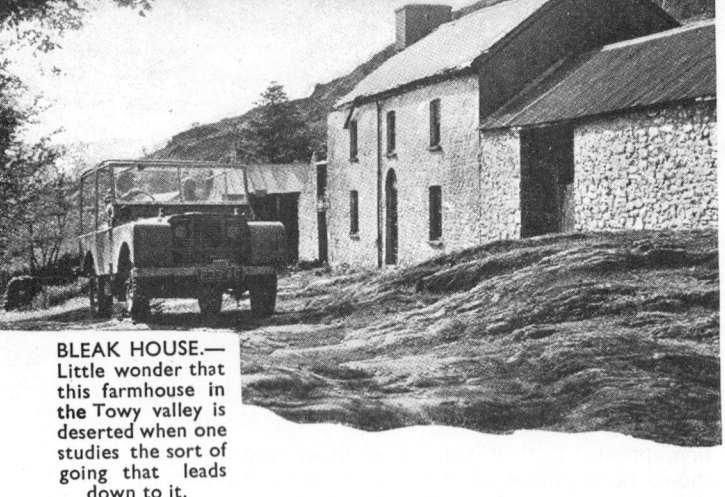

BLEAK HOUSE.—Little wonder that this farmhouse in the Towy valley is deserted when one studies the sort of going that leads down to it.

tubular hood stays, which were left on to provide additional hand-holds.

The "ancient road," known locally as the Monks' Way, leads off the old Rhayder-Devil's Bridge road just above the northernmost lake of the Elan Valley, its line being clearly defined in the form of a wide ditch reaching upwards across the flank of Esgair Rhiwlan. A guide book, written some years ago for cyclists, recommends that he who would take this way should leave his machine and proceed either afoot or on horseback.

With Lowrey at the wheel and taking our cue from what we had seen at the Rover works, we climbed the bank leading off the metalled road and started our clamber up the ditch, almost immediately to come to a standstill with spinning wheels through (a) using too low a gear and (b) showing insufficient enterprise.

The Rover transmission layout incorporates a free-wheel in the drive to the front axle, so that although the drive is permanently to all four wheels, front-wheel scrubbing is obviated. It is, however, possible, by pulling on a small ring, to lock the free-wheel when the lower ratios are in use, the lock automatically coming out of operation when the high ratios are again engaged. This device was brought into use to get us out of the slippery patch in reverse and then, in low third and showing considerably greater enterprise, we continued the ascent.

Sinking in the Ooze

A further wet-looking patch was investigated on foot, Row having in the meantime recalled reading somewhere or another that horses had been known to disappear in Welsh bogs. Having come to the conclusion that this wasn't a bog within the meaning of the act, the obstacle was attacked with encouraging success and such ease that, when the next patch of reeds appeared, we didn't even bother to prospect but, to the accompaniment of warlike whoops, charged it flat-out. This time it was a real bog, albeit not quite a bottomless

one, and the Rover sank to its chassis. Our bollarding plans came to naught because there was nothing around which to fasten the rope. The jack handle was tried as an anchor, but encountered solid rock about 9 ins. below the surface. There was nothing for it but to secure outside assistance, so while Lowrey and Ross hatched schemes which included draining the bog, Row set off on foot to a farm which could be seen on the opposite side of the lake.

Of that walk, apart from recording the fact that it was a very hot day, the less said the better. Needless to relate, there was no horse. The farmer did, however, point out another farm not a mile from where we were stuck, back on the other side of the lake but hidden from

PIN-POINTING.—A halt for photography gave an opportunity for making an exact check on navigation. The scene is half way down the Towy valley at a point where the going was reasonable.

the bog by a fold in the hills. It was a very exhausted Row who finally staggered back to the Rover with five hearty Welsh farmers to help us out.

Lowrey in the meantime had managed, with the aid of the much-bent jack handle, to lower the level of the bog by about 6 ins., find a sufficiently solid bottom to bear a jack raised on wood and stones, and move the car a good foot nearer terra-firma. With seven on the tow-rope the operation was soon complete.

Two hours' solid soaking had left the brakes completely ineffectual, but this was only discovered when taking the helpers back to their farm, sheer down the grassy hillside. Fortunately, engine compression in low second was more than adequate to hold the car.

TIGHT FIT.—Tribute to the Rover suspension is paid by this photograph in which Lowrey uses his camera over a rocky section where there was only just room to get through.

Local knowledge revealed that farther along the Monks' Way bigger and better bogs would be our portion, also that this farm was the last one for 10 miles. Three good hours had already been wasted so, with regrets, we turned towards the south and other tracks.

Forsaking the main road again at New-bridge-on-Wye and taking the by-ways to Abergwesyn, we struck off along the track towards Tregaron and then, in two or three miles, left down the valley of the Towy. There is nothing in this first stretch which could not be done by an ordinary car, and anyone caring to try it will be rewarded with some rather fine scenery. The track down the Towy is, however, a different kettle of fish. A farmer whose opinion we sought as to the possibilities of getting through obviously thought we were quite mad even to try, although he had an obviously well-used Jeep standing by the garden gate.

Having been thwarted in one enterprise, we were in no mood to fail in another through faintheartedness, and down the valley we went, finding nothing to delay our going except a narrowness of the way which at one point resulted in an argument between the centre cross-member and a solid chunk of Wales. Wales lost.

Perilous Procedure

We were now nearly down to the point where Twm Sion Catti had his hide-out, and the alternative of going straight to it or making a detour which led over a fine conglomeration of contours presented itself. Honour demanded that we choose the latter.

Local intelligence reported that only one car had ever tried even the early stages of this route, and that one had turned over; that even if we reached the bottom of the valley alive we couldn't get across the river and that, if we did, there was no road up the other side. So we set off, while intelligence moved to a vantage point from which to watch the ghastly consequences.

To the right of a narrow footpath the hillside rose almost sheer; to the left it dropped equally sharply down to the river quite enough hundreds of feet below, and it became immediately obvious that the only thing

to do was for one to walk ahead indicating to the driver where to put the near side wheels. As an added pre-caution, the doors of the Land Rover, which can be lifted off their hinges from the fully open position, were removed and put in the back, thus, as the Road Tests say, " providing easy egress."

In this fashion, and probably to the secret disappoint-ment of our solitary onlooker, we reached the bottom. What is more, after preliminary reconnaissance, we crossed the river (the Rover, incidentally, is rather good at this sort of thing). There was even a rudimentary track up through the woods on the other side. Once again lack of enterprise was our undoing, Row this time being the guilty party, worrying too much about the relative spaces between the trees and the width of the car and too little about pressing on regardless. Subse-quent investigation proved that the trees would eventually have won anyway, so turning on the hillside, which canted the Rover over at an alarming angle, back we came, through the river again and up the opposite side which, in this direction, seemed far less terrifying.

All that remained was to find Twm Sion Catti's cave, but they must have moved it since the map was made—at any rate, although we took to our feet and scrambled about like mountain goats, we couldn't find it. Nor did we see any buzzards.

(To be continued.)

UP IN THE WOODS.—When it became impossible to get through the wood for trees, the Rover had to be turned around on a steep and loamy slope, which would have been impossible without four-wheel drive.

ANY MORE FOR THE "SKYLARK"?—Return trip across the Towy river through the previously marked channel which avoided large and slippery boulders.

PART TWO

The Concluding Epi-
sodes of a Journey into
Wales in Rover's Latest
All-purpose Vehicle

HIGH DIVIDEND.—Although photographs rarely do credit to gradient, this picture does convey something of the steep-ness of Dunlop's Dividend on its upper reaches. The foot of the hill is away in the valley to the left of the car.

IT speaks well for the comfort of the Land Rover that, despite the arduous motoring we had put in during the day, we were quite ready for an after-dinner trip with Mr. and Mrs. Editor, who had come down in their own car for Sunday's attack on Dunlop's Dividend, both having been among the least unsuccessful attackers when the hill first came to notice before the war. As there were now five of us and only three upholstered seats in the Land Rover, it was the saloon which nosed its way out of Llandovery in the cool of the evening and down A.40 toward Llandilo and then on to Careg Cennen Castle, near Trapp.

At any time the old stronghold of Careg Cennen is impressive, but towards the close of day, silhouetted on its 300-ft. crag against a flaming sky, the setting is pure Dracula. Coming to the surface again after groping by fading torchlight, half double, along a winding, rock-hewn passage leading downwards from the dungeons to a subterranenan well, the hollow sound of our footsteps broken only by the steady drip of water, one almost expected to see gigantic bats flitting and squealing through the gathering dusk. It was with quite a feeling of relief that we found ourselves once again in the well-lit bar of the Castle Hotel.

Sunday morning, like its predecessor, held promise of a glorious day, and soon after breakfast off we went, in convoy this time, with Christopher Jennings at the wheel of the Land Rover—" just to see how it handles."

Not far off our route to Brechfa, nearest village to the Dividend, lies Pumpsaint, site of one of the oldest gold mines in the British Isles, started by the Ancient Britons, continued by the Romans, who had quite a settlement in these parts, and worked spasmodically until quite recently. The scene is now quite deserted. Some of the shafts, at the entrances at least, are flooded, and only piles of excavated stone testify to past activity. We spent the best part of an hour prospecting among the lumps of quartz with which the place abounds, in hopes of finding one little bit containing yellow flecks as a souvenir, but the miners have been thorough, and we pressed on to Brechfa none the richer.

The news of our coming had been bruited abroad, and

BROAD IN THE BEAM.—The way in which full advanta'e has been taken of the width of the Land Rover to seat three abreast is seen in these photographs taken at the start of the second day and along the narrow lanes to Pumpsaint.

ROVING

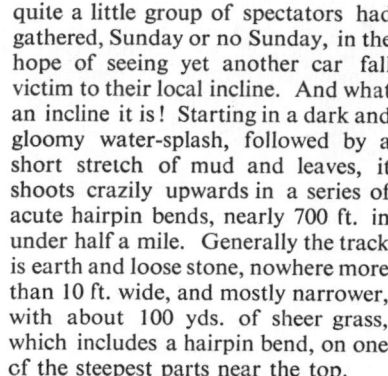

ALL ABOARD.—With nine up and an average weight of ten stone apiece, the Land Rover stormed the Dividend in terrific style with everyone hanging on for dear life.

THERE'S GOLD IN THEM THAR' HILLS.— Exploring the gold mines at Pumpsaint. Picture on right is not a new sort of voodoo dance; merely the Feature Editor prospecting for gold by cracking a piece of quartz by the simplest possible method.

quite a little group of spectators had gathered, Sunday or no Sunday, in the hope of seeing yet another car fall victim to their local incline. And what an incline it is! Starting in a dark and gloomy water-splash, followed by a short stretch of mud and leaves, it shoots crazily upwards in a series of acute hairpin bends, nearly 700 ft. in under half a mile. Generally the track is earth and loose stone, nowhere more than 10 ft. wide, and mostly narrower, with about 100 yds. of sheer grass, which includes a hairpin bend, on one of the steepest parts near the top.

It had been decided that Lowrey, as representative of technical staff, should have the honour of attempting the first ascent, and after a decent interval to allow the rest of us to reach various vantage points on foot, we heard the engine start down in the valley below, cut momentarily for the lower hairpins, and soon through the trees we caught flashes of sunlight from the screen as it came steadily upwards.

The problem of the Dividend is not only the gradient—which is ferocious enough, goodness knows !—but the acuteness of the bends, one in particular having an inner radius which cannot be greater than a couple of feet and, at the crucial point on the outer extremity, a remarkably solid stone wall, keeping back a large section of Wales. It was here that most of us were standing as Lowrey and the Rover came steadily into view : wide as he swung and hard as he locked over, the radius was too acute, and the Dividend's reputation of having never been climbed non-stop remained unbroken.

Following a couple of short reverses, a re-start was made easily. The steep grassy section might not have been there for all the notice the Rover took of it, and just to show its complete contempt, over the top it went in low top gear.

In the light of experience and upon further investigation,

ANGLING FOR POSITION.—The sharpness of THE hairpin is well shown in this photograph taken on the way down after the nine-up ascent.

it was considered that, by ploughing high up the outer bank at the beginning of the sharp hairpin—a thing which would be unfeasible with any ordinary car—it might be just possible to scrape around, so, with Row having a go this time, a further attempt was made. Like a little tank the Rover clawed its way up the bank, tilting sideways at an alarming angle, but, even by this means, there just wasn't room to get round in one sweep.

All this proved vastly intriguing to the onlookers, who had climbed this far, farming folk mostly, and naturally interested in a vehicle like the Land Rover.

It was Christopher Jennings who conceived the idea of giving them a practical taste of hill-climbing : a scheme which seemed to appeal tremendously, if one can judge by the alacrity with which five of them, plus a sheep-dog (with perhaps slightly less willingness), clambered into the back. Thus it was with nine up, hanging on to anything that afforded grip, with an average weight of about 10 stone apiece, that the Rover set off on its third climb, this time with the Editor at the wheel. On this occasion no attempt was made to take the hairpin non-stop, but rather to make the re-start under the heavy load. Except for the fact that low second, instead of low third was used for most of the climb, the vehicle might just as well have been empty—a most convincing demonstration which put the Rover well and truly on the map as far as at least five Welsh farmers are concerned. In fact, wherever we stopped with the Land Rover, it immediately became the centre of attraction—the more agricultural the country the greater the interest, and the farmers were not slow to appreciate the many practical points.

Genuinely All-purpose

What sort of car is the Land Rover? From our experience, it seems to be more or less the complete answer to the requirements of anyone who wants a genuinely utilitarian vehicle which can at the same time be used for serious everyday motoring, and the more one uses it, the more one realizes the vast amount of experience and forethought which have gone into the design.

From the very first acquaintance, it is obvious that the basic inspiration behind the design is that wartime maid-of-all-work, the Jeep. The outline is somewhat similar ; there is drive to all four wheels ; an additional set of low auxiliary gear ratios is provided, and there is the same fundamental plainness which does not sustain damage from working overalls and muddy hobnail boots. From that point the Rover goes on.

By taking fullest possible advantage of every available inch of chassis space, there is really remarkable carrying capacity. As already mentioned, three people can sit comfortably across the front, each on an individual seat which is reasonably well upholstered, so that there is no undue fatigue or soreness after a longish journey over indifferent going. Added to this, the rear portion is a practical goods-carrying space with a let-down tailboard ; or it can be used by four additional passengers, albeit without a great deal of comfort. Doors are available for the front compartment, as also are very substantial side curtains, equipped with sliding Perspex panels. The model we had was fitted with an all-over tilt, providing excellent weather protection for the complete carrying space and, very practical for the type of work which the vehicle is most likely to be used, supported on a rigid but easily-removable tubular framework.

By using as a power unit the same engine as that in the Rover " 60 " passenger car, a good degree of economy has been achieved. On the road section of the trip, with prolonged spells of fairly hard driving, a rough check showed a petrol consumption of 28 m.p.g. Moreover, the unit is smooth and willing, and while it provides no sparkling road performance, nor is such to be expected with a vehicle of this type, it cruises comfortably at 50 miles an hour on the speedometer, and, provided fair use is made of the gearbox, surprisingly good point-to-point averages can be attained.

It is no easy matter to achieve suspension which will give an easy main-road ride under light loads, and still prove satisfactory over cross-country going and with maximum weight. Yet in the case of the Land Rover, a very fair mean has been arrived at with, as is only proper, a bias in favour of load-carrying and roadless running. On the journey into Wales all three conditions were encountered. Main-road going, while not comparable to that of a modern passenger car, was by no means uncomfortable ; across country there was far less jolting than one would expect, and with the heavy load aboard on the Dividend, there were no noticeable ill-effects.

Before setting out we felt a little perturbed regarding the apparent vulnerability of the petrol tank, which is carried beneath the driving seat, but experience showed that anxiety in this regard was entirely unwarranted. Points to which we gave unqualified approval were the very large filler, enabling petrol to be poured direct from the can without spilling, and the big detachable gauze filter with which the tank was equipped.

Other details which earned full marks were the ease and speed with which the power take-off pulley attachment can be fitted or removed and the fact that the whole of the chassis, including the bumpers, are specially treated to withstand rust.

To sum up : although we failed to achieve at least three of our objectives, in no instance could the blame be laid on the Land Rover. It is doubtful whether anything short of a tank would have negotiated the morass into which we got ourselves bogged ; our failure to climb the hill near Twin Sion Catti's cave was due partly to our not having carried out a preliminary reconnaissance and largely to lack of enterprise. So far as Dunlop's Dividend is concerned, even a London taxi, had it got up that far, would not have had sufficient lock for that one hairpin.

(*We understand that since the above was written, Dunlop's Dividend has, at last, been climbed non-stop, by a specially designed, supercharged, trials car.*)

The Land-Rover, hitched to an 8-ft. ring roller and harrows, during the tests in Cornwall. A contractor estimated that six good horses would be needed to pull the implements up the 1-in-5 hill.

HARD or *Soft—*
It's all the same to the *Land-Rover*

say

L. J. COTTON, M.I.R.T.E.,
Technical Editor,
"The Commercial Motor,"

AND

T. HAMMOND CRADOCK,
Technical Editor,
"Farm Mechanization."

During 600 Miles of Road and Cross-country Work, and 12 Hours in the Fields with Plough, Harrows, Cambridge Roll and Cultivator, the Land-Rover Climbed Like a Cat and Pulled Like an Ox

NO mercy was shown during our three-day trial of the Land-Rover. It was handled ruthlessly on the road, driven "flat out" over the arduous proving grounds at Farnborough and Bagshot. and put to serious work on a farm in Cornwall. Although there was every reason for the machine to fail, it survived the rigorous tests without tiring, and without need for adjustment or replacement.

Although a small machine, it has limitless energy, which is derived from the Rover "60" engine. Because of its compactness and high output of 50 b.h.p., the engine could be well described as "potted power." Features include a three-bearing counterbalanced crankshaft, overhead inlet and side exhaust valves, hemispherical combustion space and a hydraulic automatic chain tensioner in the drive between the crankshaft and camshaft.

The four-speed gearbox, in conjunction with the two-speed transfer box, affords a selection of eight forward and two reverse ratios. Top and third gears in the main gearbox have synchronizing cones. Open propeller shafts connect the transfer box to both axles. Four-wheel drive is permanently engaged, when travelling forwards, there being a free-wheel arrangement in the front transmission to prevent "wind up" between the axles when driving on hard roads.

This free-wheel attachment isolates the drive to the front axle when reversing and a further control, which locks the free-wheel, is provided to give four-wheel drive when the reverse gear is engagd. This control will function only when the lower ratio is engaged in the transfer box, and is cancelled automatically when the lever is moved to the alternative position. Both axles are equipped with similar spiral-bevel drive and differential gears, the drive to the front axle being taken through totally enclosed constant-velocity joints.

There is nothing skimpy in construction, and the vehicle will take all punishment without flinching. Its frame is of all-welded scientifically planned construction, and is of ample proportions for any type of work. It weighs 3 cwt. more than the American Jeep, but the additional material is in the right places and is well justified.

Before starting the trials, a 12-cwt. load of sandbags and equipment was added, and the speedometer checked for accuracy at all speeds up to 50 m.p.h. The Land-Rover was left in the open at night to check ease of starting and rapid warming from cold. Starting from cold, a consumption test was made on a journey from south-east London to Farnborough. Traffic in London and suburbs was heavy, but speed was increased to 45-50 m.p.h. on the open road. These conditions were not conducive to economical performance, but no mistake was made in the quantity of fuel required to replenish the level in the tank

Climbing a 1-in-6 gradient with a water trailer containing 250 gallons of water. This was relatively easy work for the Land-Rover and low-third gear was engaged, with reserve power, before reaching the top of the incline.

Loose-surfaced gradients caused no embarrassment to the machine, which was carrying a 12-cwt. payload in addition to two observers. The Land-Rover is seen climbing a 1-in-2.74 gradient in low-second gear and the engine turning at peak revolutions.

This illustration shows the finish of changes were made at the rate of one descended at maximum speed. Wat

at the end of the course, and the consumption was equivalent to 23.5 m.p.g. The average speed, taken from the time of starting to arrival at Farnborough, was 27.3 m.p.h.

No time was wasted before tackling the steep hills at Long Valley, Farnborough. The first three loose-surfaced gradients of 1 in 3.8, 3.18 and 2.74 were climbed in rapid succession, with low-second gear engaged, and then the Land-Rover came to rest with wheelspin on the soft deposit at the foot of the next slope. This was not the limit of its climbing capacity, because it was then driven, complete with overload, up the 1-in-2¼ hard-surfaced gradient. Bottom gear was employed for this fine effort. Stop-start and reversing tests were also made on this gradient, which provided no obstacle to the machine.

Next came the "rough section," which, because of the dry, rock-hard surface, was a test of driving as well as of chassis endurance. Although a 12-cwt. load was carried, there were occasions when all four wheels were off the ground together, and the rubber buffers between the frame and axles worked overtime when the vehicle touched down.

This drastic treatment showed that adequate clearance had been left at all points on the chassis, and no weakness was discovered—except in the observer, who had difficulty in remaining in his seat! The engine was by then thickly covered with fine dust, and there was a slight leak at the joint between the air filter and induction pipe. This joint was taped as a precautionary measure.

The Land-Rover was next subjected to a 15-minute trial on the Army suspension-test course, where it was driven over two rows of kerbstones, placed with differential spacing, across the path of the vehicle. On one side, the stones were about 1 in. above ground level, whilst on the opposite side their projection was increased to 3 ins. The periodicity of the suspension was

(Above) Testing the suspension and steering at the Bagshot proving ground. The kerbstones are staggered at differential spacings and project 1-3 ins. above ground level. It was impossible to hold the steering wheel steady at 15 m.p.h.

(Left) Operating with a two-furrow plough on a loose-fallow surface. With the general-purpose bodies set to a depth of 7 ins., the Land-Rover yielded a fuel return of 6½ pints per hour during petrol-consumption trials.

(Above) Th constructio driven and arrangemer up" when

battle course at Farnborough. Gear
...s., and the gradients were climbed or
...d ditches were among the obstacles.

The Land-Rover withstood harsh treatment without faltering. It is shown riding across rough ground at high speed, and, although it carried an overload, the rear wheels were frequently off the ground. The suspension buffers worked overtime in this test.

reached at 15 m.p.h. and this speed was maintained throughout the trial.

It was only with difficulty that the steering wheel could be held, and from the outside the front axle could be seen oscillating like a fast-moving beam of a balance scale. It seemed impossible that any machine could survive this trial without breakage of steering or stub-axle assemblies. It would be difficult to find a parallel for this test in terms of road or cross-country operation.

As a finale to these tests, the Land-Rover was driven to Bagshot, where a civilian driver took the controls to show us the course—and what the vehicle could withstand. The next five miles, across water gullies, through ditches, up 1-in-2 gradients, slithering down loose-surfaced banks and many other hazards, was the most hectic drive we had ever experienced. By comparison, the trial at Long Valley was like an afternoon outing. On an average, a gear-change was made every 40 yds. It was hard work for the driver—but what punishment for the machine! We needed several seconds to regain our balance after leaving the vehicle.

The brakes were tested on the return journey between Farnborough and London, and were 100-per-cent. efficient at all speeds. The sandbags were unloaded at the works and the fuel tank replenished in readiness for the 240-mile trip to Bodmin. Little time was wasted on this journey, which was made at an average speed of 36 m.p.h., and with a fuel-consumption rate of 21-22 m.p.g. During this trip we noticed that the driving mirror required adjustment every time the door was opened and it could be better sited by attachment to the door.

On level ground, 55 m.p.h. was maintained with comfort, and there was little need to change to intermediate ratios, except for steep inclines. The highly efficient brakes and good headlamps were appreciated when crossing Bodmin Moor, where cattle roam about the road.

Although the seating is not

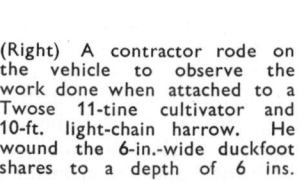

(Above) Making a stop-start test on a 1-in-2¼ concrete - surfaced gradient. Although a 12-cwt. payload was carried, the start was made without hesitation. It was necessary to engage four-wheel drive to reverse up this gradient.

(Right) A contractor rode on the vehicle to observe the work done when attached to a Twose 11-tine cultivator and 10-ft. light-chain harrow. He wound the 6-in.-wide duckfoot shares to a depth of 6 ins.

f scientific
vheels are
ree-wheel
it "wind
ard gears.

designed for 250-mile journeys, there was little discomfort experienced by the driver or passenger. A long-legged driver would appreciate more room between the pedals and steering wheel, but otherwise it was a reasonably comfortable journey. Except for short breaks, we had been travelling in the Land-Rover for over 18 hours, arriving at Bodmin in the early hours of the morning.

As we were to operate with trailed equipment on the farm, we removed the pulley drive, doors and canvas hood. The first job was to collect a filled 250-gallon water-carrier and deliver it to a meadow at the top of a 1-in-6 gradient. The trailer had a single jaw, which presented some difficulty when attaching it to the drawbar. A useful addition to the Land-Rover's equipment would be a double-jaw attachment, which could be bolted to the drawbar for use with single- or double-jaw trailers, or implements.

Low gear was selected to manœuvre out of the farmyard, the foreman walking behind with a wheel scotch in case of difficulties when climbing the hill. The vehicle showed willing, and changes to low-second and low-third were made before reaching the top.

An 8-ft. ring roller, with harrows to match, was attached for the first agricultural test, working on a light-loam soil with 2-in. tilth on granite. The throttle was set and with low gear engaged, a steady pace of 3-4 m.p.h. was maintained. The front wheels spun when attempting to operate up a steep side of the field, but with the weight of two men sitting on the bonnet, the vehicle regained traction. Normally, this tandem load of rollers and harrows is estimated to be a six-horse pull up the hill, which has a gradient of approximately 1 in 5. The speed and efficiency of the work done in this field were at least equal to that of a medium-weight tractor.

A Reasonable Load

At the next field, which was comparatively level, a Twose 11-tine cultivator and 10-ft. light-chain harrow were attached for working on a ploughed and rolled surface. The 11 duckfoot shares, 6 ins. wide, set to a depth of 6 ins., were found to be a reasonable load. With the shares in this position, the vehicle could be driven at 2 m.p.h. in super-low gear, or up to 6 m.p.h. in the next ratio.

To test the four-wheel drive, the cultivator was not lifted when turning at the headland, and it was interesting to find that the manœuvre could be made without wheel slip. In this respect, it scored over a two-wheel-drive tractor of a similar horse-power.

A fuel-consumption trial was made when the machine was attached to a Ransomes No. 3 Motrac two-furrow plough, with general-purpose bodies set to a 7-in. depth. The trial was made on loose fallow, which is notorious for lack of wheel adhesion.

At the end of the hour's fuel test, 6.3 pints of petrol were required to replenish the tank to the original level. A medium-weight wheeled tractor uses 6 pints of fuel per hour when engaged in the same class of work. We estimated that half an acre had been ploughed during the hour.

Farmers who had been sceptical of the vehicle's capabilities as a tractor began to praise it when they saw the work that had been done and our test created great interest for miles around. As an encore, we put on a special Sunday matinée for the benefit of a farm contractor who had heard of our activities. His 20-mile journey to watch the Land-Rover in the fields was not wasted.

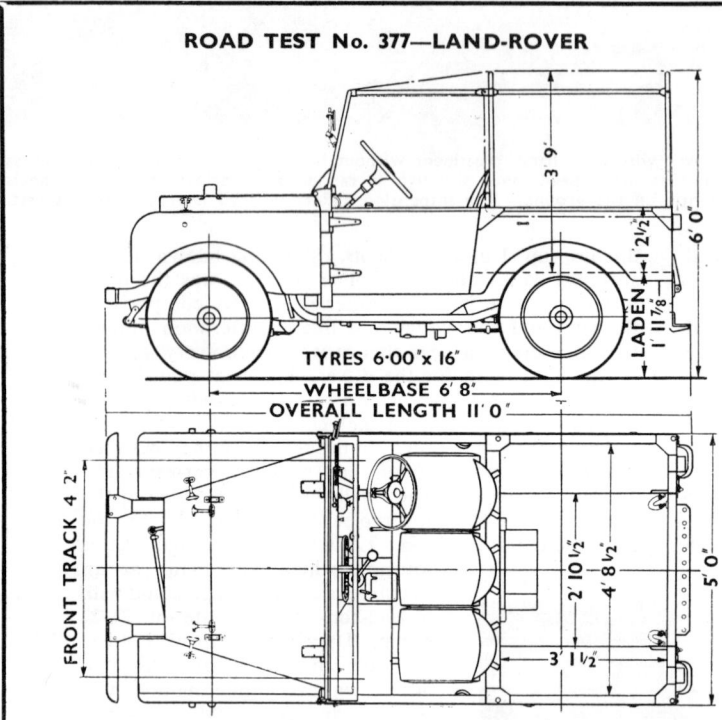

ROAD TEST No. 377—LAND-ROVER

TYRES 6·00" x 16"
WHEELBASE 6' 8"
OVERALL LENGTH 11' 0"
FRONT TRACK 4' 2"

MODEL: Land-Rover four-wheel-drive chassis.

WEIGHTS:

	Tons	Cwt.	Qrs.
Chassis and body with P.T.O.	1	4	0
Load		12	0
Driver, observer etc. ..		3	2
	1	19	2

ENGINE: Rover four-cylindered engine with overhead inlet and side exhaust valves; bore 69.5 mm.; stroke 105 mm.; swept volume 1.595 litres; maximum output 50 b.h.p. at 4,000 r.p.m.; R.A.C. rating 11.98 h.p.; maximum torque 80 lb./ft. at 2,000 r.p.m.

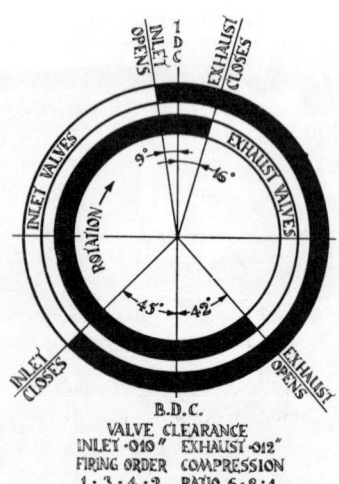

VALVE CLEARANCE
INLET ·010" EXHAUST ·012"
FIRING ORDER   COMPRESSION
1 · 3 · 4 · 2   RATIO 6·8:1

TRANSMISSION: Through single-dry-plate clutch and four-speed gearbox to transfer box, thence by Hardy Spicer shafts to spiral-bevel drives of both axles; rear axle semi-floating, front axle fully floating.

GEAR RATIOS: 2.996, 2.043, 1.49 and 1 to 1 forward; reverse 2.54 to 1; transfer box 2.888 and 1.148 to 1; axle-drive ratio 4.7 to 1.

BRAKES: Girling hydraulic to all wheels; Girling mechanical hand brake operates on transfer box output shaft; diameter of drums, 10 ins.; width of shoes 1¼ ins.; frictional area 100.8 sq. ins., that is, 51 sq. ins. per ton gross weight as tested.

FRAME: Welded box section.

STEERING: Burman worm-and-nut.

SUSPENSION: Semi-elliptic springs and Woodhead-Monroe hydraulic dampers.

ELECTRICAL: 12-volt compensated-voltage-control system and positive earth; battery 51 amp.-hr. at 10-hr. rating.

FUEL CONSUMPTION: 23.5 m.p.g. at 27.3 m.p.h. average speed; that is, 46.4 gross ton m.p.g. as tested.

TANK CAPACITY: 10 gallons; range approximately 230 miles.

ACCELERATION: 0-20 m.p.h. through gears 5 secs.; 0-30 m.p.h. 10 secs.; 0-40 m.p.h. 18 secs.; top gear acceleration, 10-30 m.p.h. 10 secs.

BRAKING: From 20 m.p.h. 13.5 ft. (32 ft. per sec. per sec.). From 30 m.p.h. 30 ft. (32 ft. per sec. per sec.).

WEIGHT RATIOS: 1.265 b.h.p. per cwt. gross weight as tested.

TURNING CIRCLE: 35 ft. both locks.

MAKER: The Rover Co. Ltd., Meteor Works, Birmingham.

Land-Rover Estate Car

Quality Production of an Essentially Practical Vehicle

ONE of the most interesting of the new vehicles which will be making their first appearance at the London Show is the Land-Rover estate car, for the inspiration of its design is sheer realism. Here is a vehicle meant for hard work, and a great variety of work. It is nearer the car of the future for the countryman than anything which has yet been produced. Names such as estate wagon and shooting brake usually denote a utility body of more or less pleasing appearance, mounted on a car chassis with perhaps some minor modifications. The Land-Rover estate car is something quite different, for it is laid out from its very foundations for the jobs it is intended to perform. It is not a special body on a touring car chassis.

The Land-Rover chassis consists of an enormously strong box-section steel frame which, in company with all exposed steel parts, is heavily galvanized to prevent corrosion. In this frame is flexibly mounted on rubber the latest Rover four-cylinder engine of 69.5×105 mm (1,595 c.c.) with the new overhead inlet and side exhaust valve arrangement and special combustion chamber, which gives high efficiency in the sense of the most power from the least consumption of fuel.

Eight Gear Ratios

This engine develops 50 brake horse power, and is cooled so that it can stand long spells of hard work. In unit with it are a large-size single-plate clutch and a four-speed gear box. At the tail of this gear box is a second transfer gear box which gives a choice of two ratios, so that the car has eight gear ratios in all. From the transfer box an open propeller-shaft runs aft to a spiral bevel rear axle, and a second propeller-shaft runs forward to a special bevel in the front axle, for this car has four-wheel drive. Drive is taken to the front wheels through a free wheel, a differential gear, and constant-velocity universal joints, all totally enclosed.

By reason of the controllable four-wheel drive, the eight speeds with overall ratios ranging from a top gear of 5.612 down to a lowest gear of 42.31 to 1, the sturdy axles, the half-elliptic springs and telescopic spring dampers, this car can be taken across country, over grass tracks, through deep water, up extremely steep slopes, and through mud with confidence, yet at the same time it is a pleasant vehicle on the main roads and able to put up a good performance.

The whole car is treated in such a way as to be able to withstand being left out in the open. In the bodywork use is made of a non-corrosive aluminium alloy sheet, and this is further treated with protective paint. The detachable disc wheels are fitted with 16×6in heavy-duty traction tyres. There is a strong bumper bar at the front, and the radiator and head lamps are protected by a strong steel screen. The front wings are detachable and a wide bonnet top gives easy access to the engine. The spare wheel is carried under a cover on top of the bonnet. Whilst no attempt at so-called styling has been made, the car has a purposeful look which is attractive.

The body is a four-square, large-capacity, design with two wide doors and a tail-board at the back opening down, together with a flap opening up, the flap being mostly rear window. The body is the full width of the vehicle. When used for carrying passengers it can accommodate seven people, for there are three separate bucket seats in front, and side seats for four in the back.

The flanking front seats can be folded and tipped up right out of the way to give access to the rear seats through the doors. The rear seat cushions can be removed entirely if the back compartment is needed for luggage or goods. Lastly the car is arranged so that it can tow a trailer, and is able to exert a draw-bar pull of 1,200 to 1,800 lb. A more sensible practical vehicle of many uses would be difficult to envisage, added to which the Land-Rover is as well made as it is well designed. The main dimensions are: Wheelbase, 6ft 8in, track 4ft 2in; overall length, 11ft; width, 5ft; approximate weight 27cwt. Twelve-volt electrical equipment is fitted, including a 52 ampère-hour battery. The fuel tank capacity is ten gallons.

Land-Rover Stops the Home Fires Burning

Dashing through a cornfield on its way to a fire, the Land-Rover reveals complete disregard of conditions underwheel.

FIRE fighting in remote country districts has its own peculiar difficulties brought about by the absence of metalled roads, inadequate water supplies and by high risks. Everything militates against speedy and effective action. The conventional, heavy, fire-fighting appliance does not take too kindly to farm tracks. In such areas a small, highly manœuvrable and fast appliance is required to reach the fire before it has got out of hand.

The specially adapted Land-Rover, demonstrated at Laindon, Essex, last week, by Essex County Fire Brigade, meets the case admirably. Fitted with a reel carrying 180 ft. of rubber hose, a 16-ft. light-alloy extension ladder, a searchlight and a bell, the Land-Rover becomes a versatile fire-engine. A 40-gallon water tank enables the crew of two to deal with a fire for 8½ minutes,

during which time it would be possible for a larger appliance to approach and lay down the required length of hose.

The appliance also carries soda-acid, foam and C.T.C. extinguishers, first-aid gear, stirrup pump, canvas buckets and lengths of canvas hose. The standard power take-off has been adapted to drive a gear-type pump, supplying the hose at a maximum pressure of 100 lb. per square in. Less than 6 b.h.p. is required of the 55 b.h.p. engine to drive the pump.

The equipment, stowed in and on the machine, weighs approximately 12 cwt., and thus the outfit is somewhat overloaded. The rear springs have been strengthened and the tyre pressures raised to deal with the weight.

The appliance was driven to a typical estate within the Laindon Brigade area, the driver remarking on the way that

since being put into commission on July 29, the Land-Rover had attended 13 fires, seven of which could not be approached by any other appliance. It was soon seen that this was no understatement. The estate in question appeared to have been built with little regard to the question of accessibility, the so-called approach road resembling a model of a mountain canyon.

Level in parts, but with steep gradients as well, it was rutted, fissured and beset with hazards of all kinds.

The appliance descended a steep hill, and the hose was run out to show the pump in action.

No sooner had this been done than a train passed on a nearby line and a spark ignited the grass embankment. This opportunity was not missed, and the fireman ran his hose rapidly down to the conflagration and within 30 secs. the fire was extinguished.

Rough or Smooth

In approaching the houses on the estate, most of which were wood bungalows, the appliance charged through ditches, hedges, ploughed fields and other natural hazards, only once failing to reach its objective in the gear selected. On that occasion the low ratio box had to be brought into use.

A representative of " The Commercial Motor " who drove over the demonstration route, found that there was plenty of power in hand for fast journeys, 40 m.p.h. being reached quickly, although it was obvious, on changing into top gear, that the load carried was heavy. The brakes required the minimum of effort to stop the laden appliance easily under adverse conditions.

Officers of visiting fire brigades were impressed by the demonstration, at the conclusion of which orders for six more Land-Rovers were being discussed. The first machine, supplied by Spurling Motor Bodies, Ltd., Colchester, was adapted at the brigade workshops at Colchester by Mr. C. G. Sayer, the chief mechanical engineer, and a second is soon to go into operation. The six new vehicles would probably be fitted with radio and used as fire-control cars.

Land-Rovers Cover 22,000 Tough Miles

AFTER five months in Africa and the Middle East, a party of four Land-Rovers led by Col. LeBlanc has returned to this country. The vehicles have covered 22,000 miles over tough country and in extremes of temperature.

The manufacturer's object in this expedition was to put the vehicles through the most searching of African

routes at a time of year deliberately chosen for their difficulty, and to collect technical data on performance under these conditions.

An average of over 190 miles was covered each running day and the Land-Rovers experienced no involuntary stops, even in sections which had never been thought possible for vehicle travel. The most difficult sections were Timbuctoo to Bourein over loose sand and rock, which had to be covered as a

detour round floods; the Southern Egyptian desert from Port Sudan to Asuan where the temperature was 120° in soft sand; and Baghdad through the desert to Damascus—nearly 600 miles of sandstorms, and 124° temperature.

A high-speed run covered the Luxor-Cairo section of 425 miles of dirt roads in only one long day. Returning through Jugoslavia, the 213-mile Belgrade-Zagreb section was covered in four hours' running time.

Land-Rover
1948-50 Models

Articles in this series are written by the Technical Staff of "The Motor Trader" and checked by the service managers of the vehicle manufacturers or importers.

Manufacturers: Rover Co., Ltd., Solihull, Birmingham.

FIRST produced in August, 1948, this vehicle was designed as a general purpose utility car primarily for farmers. Briefly, it uses the same engine as the 1948 Rover 60 car and a similar gearbox, modified to incorporate a two-speed transfer gearbox for the four wheel drive. Front and rear final drive units, which are interchangeable, are the same as those on the car. The front wheel drive is permanently engaged through the free wheel.

Vehicle numbers consist of a six-, seven- or eight-figure number prefixed R or L to indicate right- or left-hand drive. Originally the first figure (8) indicated the year of manufacture (1948), though this was carried on through the 1949 season. The second figure (6) indicated Land-Rover as distinct from car. The last four figures were the serial number, starting from 0001. Later the station wagon was introduced and was given serial numbers starting from 70001, so that the whole number appeared as 8670001. For the 1950 season a five-figure serial number is used throughout, and a third prefix figure is added to indicate the type of vehicle. The basic vehicle is 1, the station wagon 2 and the welding outfit 3. The vehicle number is stamped on a plate on the near side of the dash under the bonnet. The chassis number is the same, and is stamped on the nearside front engine mounting bracket. The engine number, which does not necessarily correspond, is on the nearside of the cylinder block.

All threads and hexagons are B.S.F.

Instruments and controls:

1. Panel lamp switch
2. Lighting and ignition switch
3. Choke control warning light
4. Ammeter
5. Oil pressure warning light
6. Petrol gauge
7. Speedometer
8. Choke control
9. Screenwiper switch
10. Horn push
11. Dipper switch
12. Clutch pedal
13. Accelerator
14. Brake pedal
15. Handbrake
16. Freewheel lock
17. Gear lever
18. Transfer lever
19. Hand throttle
20. Ignition warning light
21. Starter switch
22. Inspection lamp sockets

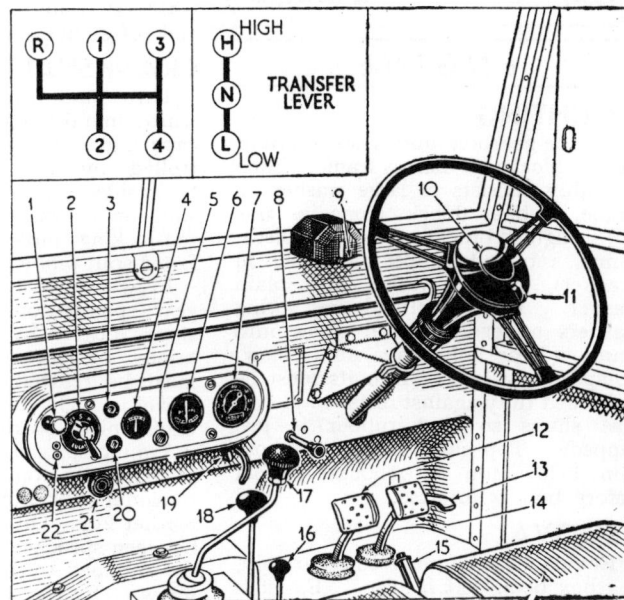

ENGINEERING CHANGES

ENGINE	Engine No.	VEHICLE	Vehicle No.
Carburettor linkage adapted for fitting of governor	861872	Brake fluid tank moved from scuttle to seat box	861001
Starter changed from M418 G/C to M418 G/76	862235	Gear lever moved from cover panel to gearbox casing. Pressed dash adopted. Seat box and floorboards changed	861501
Main bearing nuts changed from self-locking to castellated	867432		
Clutch assembly changed from Rover to Borg & Beck. Release sleeve changed from built-up to integral...	06100201	Front shock absorbers changed to same type as rear. Extra leaf added to front springs	862115
Dynamo changed from C39P to C39PV	06106001	Rear spring camber (o/s only) changed	862298
Clutch release sleeve changed to slide in bush in housing, was bushed on primary shaft. Cross-shaft serrated, was splined	06106001	Rear spring camber (both) changed ...	8664115
		Hydrastatic brakes discontinued—Snail cam adjustment adopted, smaller wheel cylinders ...	8667721
AXLES (FINAL DRIVE)	**Axle No.**	Front bumper brackets welded to bumpers instead of frame ...	06100156
Ratios changed from 4.88 to 4.7 (approx. unit no.)	861371	Freewheel control changed—Push knob instead of pull ring. Offside floorboard changed	06104001
Bevel pinion shaft bearings changed from ball and roller to taper roller. Shorter pinion shafts, longer prop. shafts...	06106001	Control box changed from RF96 to RF95. Fuse box discontinued ...	06106001

DISTINGUISHING FEATURES—No outward changes have been made during the currency of this model except that rope cleats have replaced strap loops for the hood fixing. Note alternative mounting of spare wheel

ENGINE DATA	
No. of cylinders	4
Bore × stroke : mm	69.5 × 105
in	2.74 × 4.10
Capacity : c.c.	1595
cu. in	97.3
R.A.C. rated h.p.	11.98
Max. b.h.p. at r.p.m. ...	50 at 4,000
Max. torque (lb/ft) at r.p.m....	80 at 2,000
Max. drawbar pull ...	1,200—2,000 lb
Compression ratio ...	6.8 : 1
Compression pressure (300 r.p.m.)	125–135 lb/sq. in

ENGINE

MOUNTING

Engine-gearbox unit rests on four rubber blocks on chassis frame. Each assembly consists of large washer on frame, rubber block (concave side up), engine bracket, distance-piece, shims, rebound rubber (spigoted in bracket), rubber washer and plain washer. Front bolts have spring washers under head and screw into frame. Rear bolts have nuts below, no spring washers. All bolts must be tightened fully against distance-pieces. Use shims so that rubber is just nipped. Top shim should be about $\frac{1}{8}$ in below top of rebound rubber before bolt is inserted.

REMOVAL

Engine must be removed separately, gearbox being left in place. Easier if sump is removed. Remove bonnet top by opening fully and sliding off hinges. Detach radiator grille, disconnect headlamp wiring from junction box on dash and clips on wing, disconnect radiator hoses and remove grille panel and radiator core assembly (3 bolts each side, 3 at bottom with square rubber packing pieces). Disconnect all pipes, wires and controls, and remove air cleaner. Take weight of engine on slings below crankshaft pulley and between sump and flywheel housing, remove front mountings and support front of gearbox on jack. Take off 13 nuts (plain washers) round bell-housing flange, and lift engine forwards and upwards.

CRANKSHAFT

Three main bearings. Steel-backed, white metal-lined shells dowelled in caps and crankcase. End float controlled by centre bearing, flanged both sides. No hand fitting permissible on bearings. Caps retained by self-locking nuts on earlier engines. Later split-pinned.

Flywheel, with integral starter ring gear, spigoted on rear flange of crankshaft, located by two dowels and retained by six setscrews. Flywheel must be refitted with "O" marked dowel in "O" marked hole in wheel. Oilite spigot bearing bush in flywheel.

Timing sprocket keyed on front end of shaft with combined fan pulley and torsional vibration damper by single feather key. Oil thrower disc trapped between sprocket and pulley. Assembly retained by hand starter dog setscrew. Pulley hub passes through lipped oil seal in timing cover.

NUT TIGHTENING TORQUE DATA	
Main bearings	120 lb/ft
Cylinder head	45–50 lb/ft

CRANKSHAFT DATA				
	Main Bearings			Crank-pins
	No. 1	No. 2	No. 3	
Diameter	2 in	2 in	2 in	$1\frac{7}{8}$ in
Length ...	$1\frac{13}{16}$ in	2 in	$1\frac{13}{16}$ in	$1\frac{3}{8}$ in
Running clearance :				
main bearings			.001–.002 in	
big ends			.001–.002 in	
End float :				
main bearings			.003–.005 in	
big ends			.009–.013 in	
Undersizes			.015, .030, .040 in	
No. of teeth on starter ring				
gear/pinion			97/11	
Con. rod centres			$7\frac{1}{2}$ in	

Rear main bearing cap fits in square recess in crankcase with square cork seals in grooves. Fit seals in cap before inserting cap. Split oil collector housing fits round rear of shaft, and is bolted to rear face of crankcase and cap.

CONNECTING RODS

Big ends thin wall, steel-backed, white metal-lined shells located by tabs. No hand fitting permissible. Small ends bronze bushed.

Fit piston and rod assembly with bleed hole in shoulder of big end away from camshaft.

PISTONS

Aluminium alloy, solid skirt, oval ground. Gudgeon pins located by spring rings. Fit of pin in piston is critical. Pin must not fall through by its own weight, but push required to insert it must not exceed 100 lb at 50-70 deg. F.

Compression rings are taper faced. Fit with side marked "T" upwards.

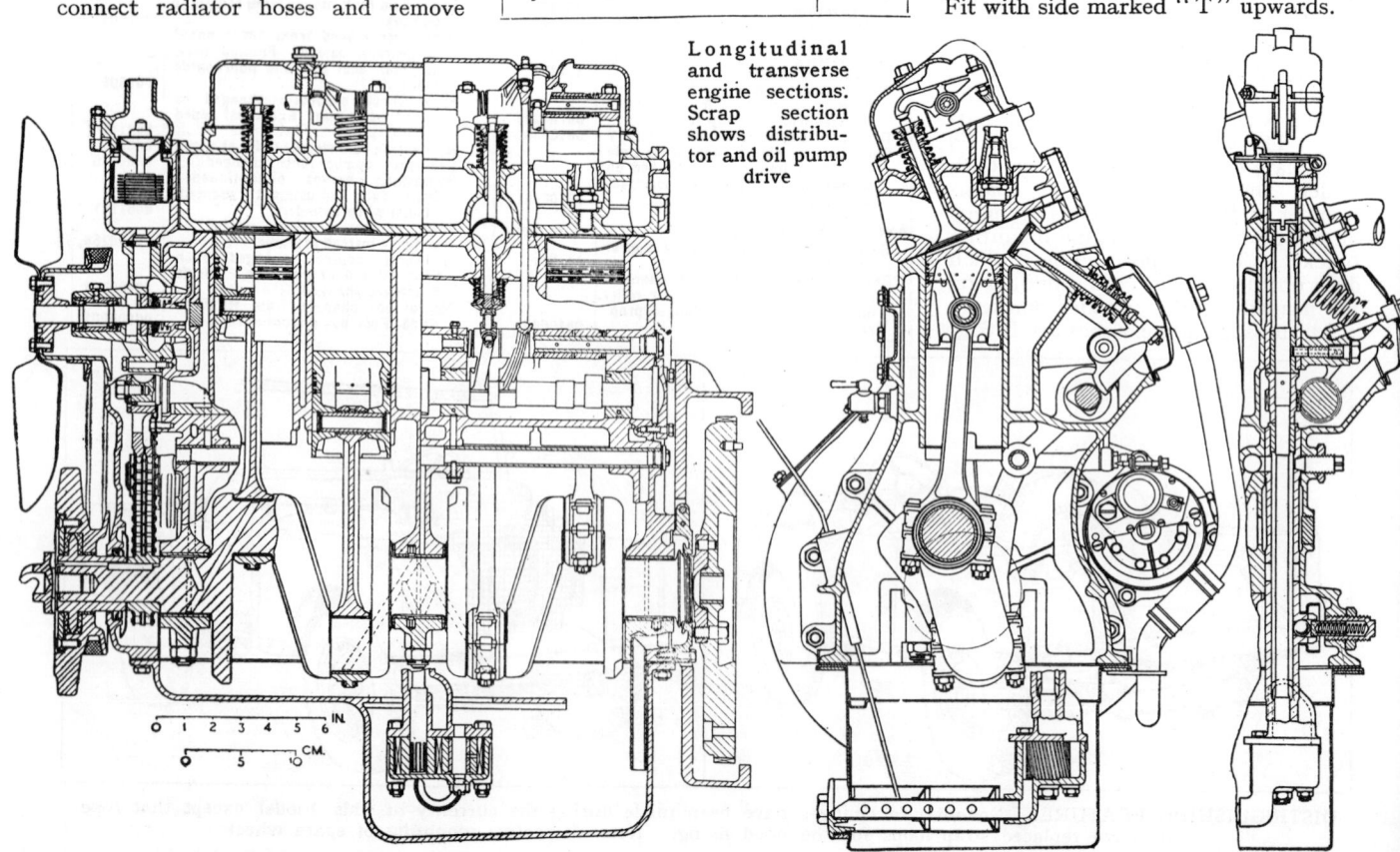

Longitudinal and transverse engine sections. Scrap section shows distributor and oil pump drive

Big ends will not pass through cylinders, nor will pistons pass crankshaft. Push con. rod up as far as it will go, and turn 90 deg., when it will drop into slots in cylinder skirt, and gudgeon pin can be pushed out.

PISTON DATA		
Clearance (skirt, major axis) ...	.0015 in	
Oversizes ...	.010, .020, .030, .040 in	
Weight (with rings and pin)	14oz 8dr	
Gudgeon pin : diameter ...	$\frac{11}{16}$ in	
fit in piston ...	Hand push (see text)	
fit in rod ...	Easy sliding	
Compression height (peak of crown) ...	2.534 in	
	Compression	**Oil Control**
No. of rings ...	2	2
Gap ...	.014–.018 in	.011–.015 in
Side clearance in groove	.0005–.002 in	.0005–.002 in
Width of ring...	.070 in	$\frac{3}{16}$ in

CAMSHAFT

Duplex roller endless chain drive with jockey sprocket tensioner.

Camshaft sprocket keyed on shaft with Woodruff key and retained by setscrew and large washer. Shaft runs in four split Mazak bearings, halves dowelled together and located by setscrews accessible inside lower rocker cover. End float controlled by thrust plate between sprocket and shoulder on shaft, bolted to crankcase.

Tensioner consists of jockey sprocket on swinging arm with ratchet at free end in which pawl on crankcase engages. Tension provided by cylinder with spring-loaded piston which is located in recess arm. Oil pressure from lubrication system augments spring, and oil is trapped by non-return valve in base of cylinder to give hydraulic lock. Spring rings retain jockey sprocket on arm and arm on pivot pin.

Camshaft can be removed with engine in place. Draw off torsional vibration damper and remove timing cover. Extract split pin and pull off hydraulic tensioner, piston and spring (carefully, as spring is very long when released). Release pawl and remove jockey sprocket and arm assembly. Chain will then be slack enough to be lifted off. Draw off camshaft sprocket, which has two holes tapped $\frac{3}{8}$in Whit. for withdrawal. Remove distributor, housing and long vertical shaft with skew gear as assembly. Extract lower rocker shaft, take out bearing locating setscrews and draw out shaft with bearings, picking out halves.

To retime valves with timing chain and tensioner off, set exhaust tappets to running clearance, slacken inlet tappet screws right off and turn camshaft in running direction until No. 1 exhaust valve is fully open (use dial indicator if possible). Turn crankshaft in running direction until EP mark on flywheel is opposite pointer (visible under trap on off side of flywheel housing). Assemble timing chain so that there is no slack on driving side, and fit jockey sprocket assembly. Check timing. Camshaft sprocket has three keyways for fine adjustment. When timing is right, fit tensioner cylinder, piston and spring, dry to prevent formation of air lock.

CAMSHAFT DATA		
	No. 1	**Nos. 2, 3, and 4**
Bearing journal : diameter	$\frac{7}{8}$ in	$\frac{7}{8}$ in
length	$1\frac{3}{8}$ in	$1\frac{1}{16}$ in
Bearing clearance ...	.001–.0025 in	
End float ...	.003–.005 in	
Timing chain : pitch	$\frac{3}{8}$ in	
No. of pitches ...	78	

VALVES

Inlet in head, exhaust at side. Not interchangeable.

Split cone cotter fixing, double springs. Inner springs (right-hand coil) fit tightly in outer springs (left-hand coil) by selective assembly, and should always be replaced in pairs. Inlet valve spring collars have synthetic rubber ring inset at bottom and sealing on stem. Ring is difficult to see unless looked for carefully. On latest engines rubber rings are in top of inlet valve guide instead of in collar.

Valves guides shouldered, not interchangeable, exhaust longer than inlet. Exhaust valve seats pressed in. Break with chisel to extract for renewal. Material is very brittle and apt to fly. Take care and hold thick pad over work to catch fragments.

VALVE DATA		
	Inlet	**Exhaust**
Head diameter ...	$1\frac{17}{32}$in	$1\frac{11}{32}$in
Stem diameter ...	$\frac{5}{16}$in	$\frac{3}{8}$in
Face angle ...	30 deg.	45 deg.
Tappet clearance (cold)	.010in	.012in
	Inner	**Outer**
Spring length : free ...	1.817in	1.845in
fitted	1.469in	1.625in
at load ...	10 lb	30 lb

ROCKERS

Two sets of rockers. Lower set act as cam followers, operating exhaust valves directly and inlet valves through push rods. These rockers are of cast iron, and bear directly on shaft, with bronze thrust washer between each pair.

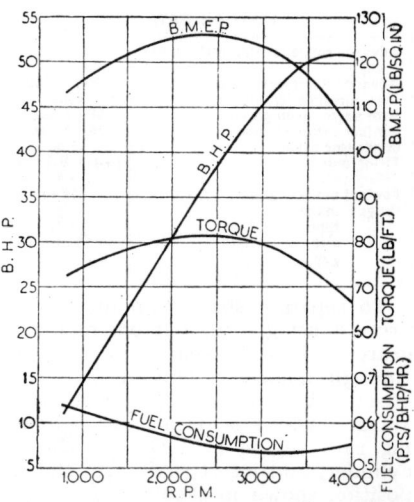

Inlet rockers on head are bushed, offset in pairs left- and right-hand, and carried on single hollow shaft located by setscrew in rear pillar, which is drilled for oil feed. Springs fit outside each pair of rockers.

Lower rocker shaft is in two pieces, inserted from front and rear and located by setscrews. Shafts can be extracted from rear with engine in place, through trap in dash (on earlier vehicles hole may have to be drilled). Unscrew rear plug and draw out rear shaft with extractor, which is long enough to reach front shaft.

LUBRICATION

Gear pump in sump, spigoted in crankcase by integral drive housing and located by grub screw and locknut outside. To remove pump first extract relief valve plunger and spring (large grub screw and locknut below locating screw). Then slacken square-ended grub screw above relief valve, and draw out pump, leaving shaft (splined in pump gear) in place.

Cylindrical gauze suction strainer screwed into sump and registering with pump intake. Must be extracted before sump is removed.

AC bypass pressure filter type ZS1.

Ball relief valve located in side of pump drive housing, with spring and plunger inserted from outside, adjustable by grub screw and locknut. Normal pressure 40lb at 30 m.p.h. If oil cooler is fitted, special relief valve spring $4\frac{1}{2}$in long is supplied, which boosts pressure to 75-80lb.

IGNITION

Anti-clockwise distributor, with centrifugal and vacuum control and micrometer adjustment, spigoted in drive housing and held by setscrew.

Drive housing spigoted in cylinder block and located by hollow setscrew inside lower rocker cover. Long drive shaft driving oil pump at lower end, has skew gear pinned to centre (larger boss upwards), and is supported on either side of gear by flanged bronze bushes, one in pump drive housing and other in distributor drive housing.

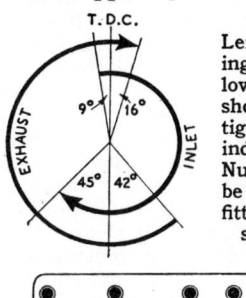

Left : Valve timing diagram. Below : Diagram showing order of tightening of cylinder head nuts. Nuts marked X to be tightened after fitting of rocker shaft brackets

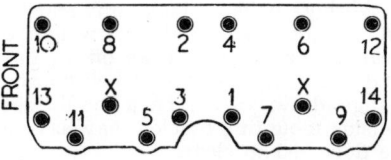

IGNITION DATA	
Advance range (centrifugal) (crank deg.)	36—40 deg.
Advance starts (crank r.p.m.)	1200—1600
Max. advance (crank r.p.m.) ...	4600
Cam angle (open period) ...	41 ± 4 deg.
Contact spring tension	20—24 oz.
Condenser capacity	.2 mf
Firing point	15 deg. B.T.D.C.
Firing order	1 3 4 2
Contact breaker gap	.012 in
Plugs : make	Lodge
type	HLNR
size	14 mm
gap...	.023—.026in

To separate shaft from distributor drive housing drive out pin retaining gear, and press shaft out of gear through housing. Driving dog with offset slot pinned to top end of shaft. When refitting shaft and drive housing turn engine to T.D.C. on No. 1 cylinder, and insert shaft with slot in position shown in sketch.

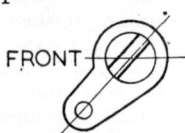

Set contact points to break 15 deg (four flywheel teeth) before T.D.C. FA mark on flywheel. Set octane selector on distributor at long line on scale.

COOLING SYSTEM

Pump, fan and non-adjustable thermostat in housing bolted to cylinder head above pump. Pump has spring-loaded carbon and rubber seal.

Pump can be removed with radiator in place. Remove thermostat housing first, noting rubber ring between housing and pump. Then take off fan blades and pulley.

To dismantle pump draw off impeller, which is pressed on shaft and has two holes tapped in ⅛in B.S.F. for extractor. Behind impeller on 1948-9 pumps are spring, brass retainer, rubber seal and carbon disc. Latest pumps have carbon and rubber bonded seal unit in unslotted impeller.

FUEL SYSTEM DATA		
Carburettor : make		Solex
type		32 PBI—2
Settings : choke tube		23
main jet		107.5
correction jet		160
idling jet		45
pump or speed jet ...		50
cruising (economy) jet ...		50
idling air bleed		1.5
starter petrol jet (GS) ...		135
starter air jet (GA) ...		5.5
needle valve		1.5
Air cleaner : make		AC
type : oil bath ...		1574191
centrifugal ...		1579219
Fuel pump : make		S.U.
type		L
pressure		1 lb/sq. in

Take out bearing locating setscrew and push out shaft and ball bearing assembly. Replacement shafts supplied with bearing and pulley flange.

When assembled, impeller should be pressed on until there is .020in clearance between vanes and housing. If possible latest seal and impeller assembly should be fitted.

Adjust fan belt by swinging dynamo until there is about ½in movement either way on longest run of belt.

TRANSMISSION

CLUTCH

Single dry plate. Rover make with Borg & Beck driven plate on early vehicles, later complete Borg & Beck assembly. No internal adjustment on Rover type. Journal ball release bearing enclosed in separate housing bolted to gearbox, from which it is lubricated, and operating clutch fingers through sliding sleeve.

Only adjustment is by nut on rear end of pedal pull rod, to give ¾in free movement at pedal pad.

Gearbox can be removed for service to clutch without disturbing engine.

GEARBOX

Four-speed, synchromesh on top and 3rd gears, single helical constant mesh gears except for 1st and reverse. Two-speed transfer box bolted to rear.

To remove gearbox and transfer box, disconnect battery, remove hood

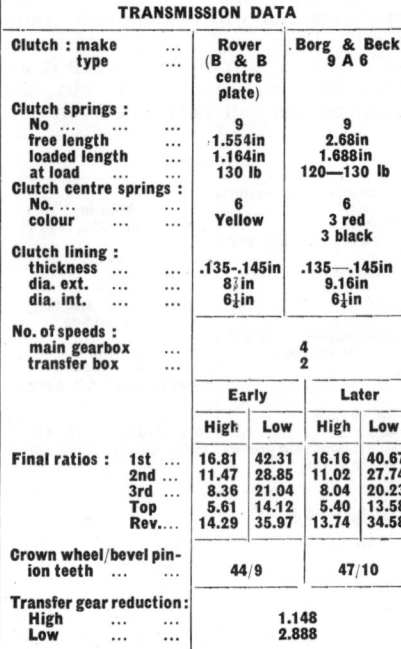

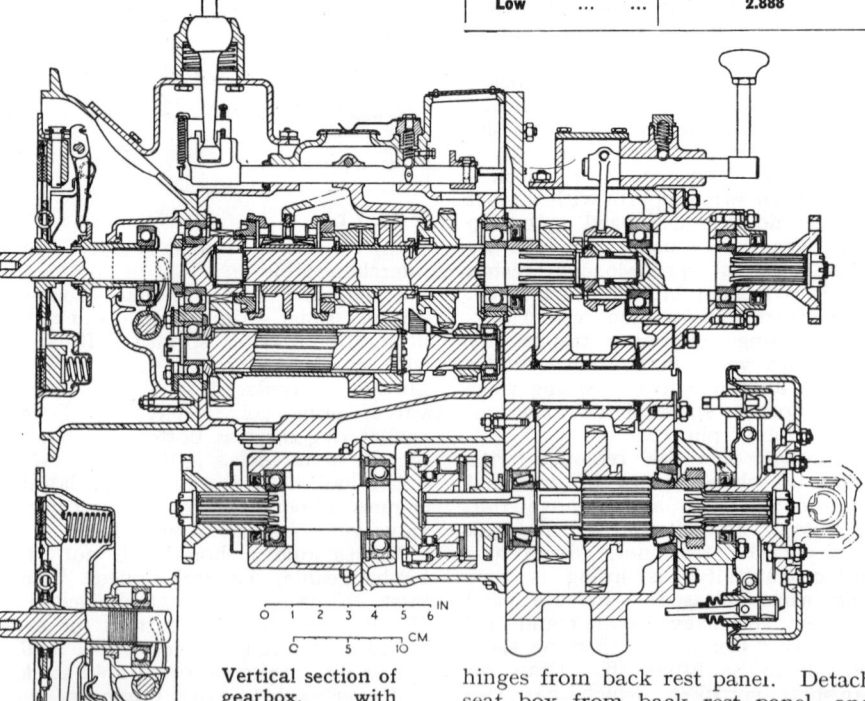

Vertical section of gearbox, with transfer box and free wheel sectioned in plane of main shafts. Below, left : Borg & Beck clutch and latest withdrawal assembly

(if fitted), detach centre panel from seat box, disconnect handbrake rod from bell-crank and remove lever assembly complete, drawing end of lever back through draft excluder. Unscrew knobs from gear lever, transfer lever and freewheel control (earlier ring). Take off clutch and brake pedal pads. Take up floor plates and remove gearbox cover. (On early vehicles gear lever will come away with cover.) Detach lids of petrol tank and tool box by removing

TRANSMISSION DATA				
Clutch : make ...		Rover		Borg & Beck
type ...		(B & B centre plate)		9 A 6
Clutch springs :				
No		9		9
free length ...		1.554in		2.68in
loaded length ...		1.164in		1.688in
at load ...		130 lb		120—130 lb
Clutch centre springs :				
No.		6		6
colour ...		Yellow		3 red
				3 black
Clutch lining :				
thickness ...		.135-.145in		.135—.145in
dia. ext. ...		8⅜in		9.16in
dia. int. ...		6¼in		6¼in
No. of speeds :				
main gearbox ...			4	
transfer box ...			2	

		Early		Later	
		High	Low	High	Low
Final ratios :	1st ...	16.81	42.31	16.16	40.67
	2nd ...	11.47	28.85	11.02	27.74
	3rd ...	8.36	21.04	8.04	20.23
	Top ...	5.61	14.12	5.40	13.58
	Rev....	14.29	35.97	13.74	34.58

Crown wheel/bevel pinion teeth		44/9	47/10
Transfer gear reduction:			
High		1.148	
Low		2.888	

hinges from back rest panel. Detach seat box from back rest panel, and floor sills from dash. Lift off seat box.

Remove rear propeller shaft complete, and disconnect front shaft and p.t.o. shaft (if fitted) at gearbox end. Disconnect clutch linkage, unhooking return spring from front end first. On right-hand drive vehicles cross-shaft is jointed to extension shaft, supported in spherical bush on side member. Extract either joint pin (split pinned) and push extension shaft clear. Disconnect speedo drive and remove rear mounting bolts. Jack up rear of engine about ½in, and take weight of gearbox on slings. Take off 13 nuts and plain washers round bell-housing flange, draw gearbox back and lift out. Offside mounting bracket may have to be detached to clear.

To remove transfer box detach main gear lever assembly and reverse stop (later models) from gearbox and bell-housing. Take off transfer cover plate (with plug) on top of transfer box. Detach freewheel control rocker and unscrew eyebolt. Detach transfer gear lever with link from selector rod and bracket. Remove lever and detach bracket from bell-housing. Detach freewheel control lever assembly. Transfer box with freewheel and output housing assembly can then be separated from gearbox. Remove transmission brake drum and draw off rear driving flange. Detach brake backplate assembly and bottom cover. Remove idler spindle locking plate and push spindle out to rear, catching idler gear. This gives access to three nuts inside which, with six outside, hold transfer box to gearbox. Power take-off drive housing or blanking housing, with roller spigot bearing, should first be removed from back of transfer box.

To dismantle gearbox follow procedure in "Trader" Service Data sheet No. 150, covering Rover 60 and 75, which have similar gearbox.

To reassemble gearbox reverse order of dismantling, observing following :—

End float of 2nd and 3rd gears should be .002-.004in. Thrust washers .125, .128, .130 and .135in thick.

Conical distance-piece for front end of layshaft available in .312, .332 and .352in thicknesses to take up end float.

Note that felt sealing washers fit on 1st/2nd and 3rd/top selector rods, cork washer on reverse rod.

When inserting selector springs, note that 3rd/top (nearside) and 1st/2nd (centre) springs are same, but reverse (offside) is stronger.

Adjust 2nd speed stop screw so that with 2nd gear engaged there is .002in clearance between screw head and stop on selector rod.

TRANSFER BOX

Idler gear cluster runs on caged roller bearings on spindle, with thrust washers at both ends, tabbed to locate in casing.

Output shaft carries constant mesh high gear (free) and sliding low gear (splined) between taper roller bearings, adjusted by shims (.005, .010, .015in thick) between casing and speedo drive housing.

Rear extension of output shaft carries speedo drive gear, nipped between inner race of rear bearing and driving flange, which also carries brake drum.

Forward extension of output shaft carries sliding dog for locking freewheel, and hub of freewheel. Outer member of freewheel bolted and dowelled to register on flange of forward output shaft, which is carried in ball bearings in freewheel housing and output housing, and has driving flange on outer end.

On early vehicles forward output shaft was in two parts, connected by dog coupling. Inner part, carrying female member of coupling, bushed for spigot of outer part.

To remove freewheel assembly with gearbox in chassis, remove freewheel control, rocker and eyebolt, transfer lever assembly and forward propeller shaft. Draw off driving flange and draw out freewheel control rod. Remove plug on inner side of output housing with transfer selector ball and spring. Detach output housing with shaft and, on early vehicles, locking dog. Take out freewheel housing nuts and slacken rear engine mountings, levering gearbox up to allow freewheel housing to clear frame as it is drawn forward. Draw spring guide and freewheel operating spring from transfer casing.

When refitting output housing, cover end of transfer selector rod with thimble to protect oil seal in housing.

To dismantle freewheel take out locking dog, fork and sleeve. Extract spring ring retaining freewheel driven member in bearing, and press shaft out of ball bearing. Extract outer spring ring and press bearing out of housing.

Extract spring ring and splined retaining plate inside freewheel drum, and draw out hub with three sets of three graded rollers, blocks and springs. These parts should be marked with paint or carefully placed so that they may be refitted in same relative positions. Freewheel hub runs in caged roller bearing in drum, which is detachable from shaft.

To dismantle transfer box after removal from gearbox, with brake drum, driving flange, freewheel and idler gear off, remove speedo drive housing and shims. Pull off speedo drive gear. Tap output shaft back until outer race of rear bearing is free of casing. Extract spring ring retaining outer race of front bearing and tap output shaft forward as far as possible. Slide shaft back and insert aluminium packing pieces between rollers and outer race. Drive shaft forward again and repeat if necessary with thicker packings until outer race is free. Draw off inner race. Spring ring retaining high transfer gear on shaft with thrust washer can then be extracted, and shaft pushed out to rear through gears, which will drop out.

To reassemble transfer box reverse order of dismantling, observing following points :—

End float of high transfer gear on shaft should be .006-.010in. Grind thrust washer if necessary.

Output shaft bearings should be adjusted by shims between transfer casing and speedo drive housing to be free without play.

Transfer selector fork should have thread for pinch-bolt towards nearside.

Idler gear cluster should have .005-.010in end float.

PROPELLER SHAFTS

Hardy Spicer needle roller bearing universal joints, series 1300 for both front and rear drive shafts and for p.t.o. shaft if fitted. Nipples for lubrication of joints on later vehicles.

REAR AXLE

Semi-floating, spiral bevel drive. Rear cover welded to banjo casing.

To remove axle from car either drop rear ends of springs and roll axle back, or remove half-shaft and brake assemblies and final drive, and pass out sideways through springs.

Axle details are similar to those of 1948-9 Rover 60 and 75 cars. See "Trader" Service Data sheet No. 150.

On latest axle bevel pinion shaft is carried in taper roller bearings, adjusted by shims (.003, .005, .010in thick) between inner race of front bearing and shoulder on shaft, to give preload of 10 lb/in. Mesh adjusted by shims behind outer race of rear bearing (same thicknesses as bearing shims).

Crown wheel spigoted and attached to flange of one-piece differential cage by two fitted bolts and eight set-

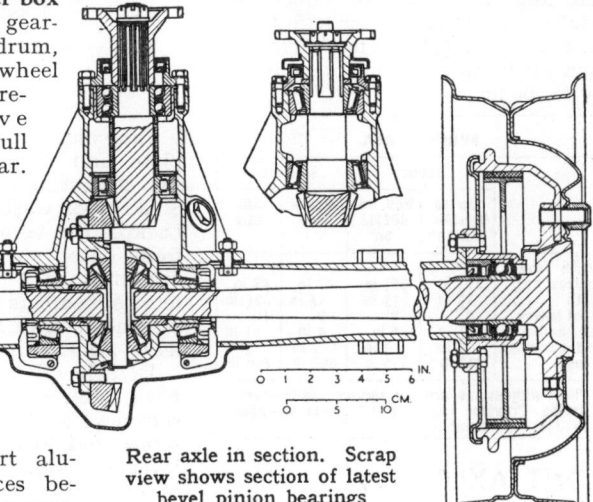

Rear axle in section. Scrap view shows section of latest bevel pinion bearings

screws. Side bevel gears have fibre thrust washers behind, .040in, .045in or .050in thick. Differential assembly carried in taper roller bearings in split housings, with ring nuts to adjust bearings and mesh. Bearings should be preloaded .005in (one serration on nuts), and backlash should be .007in, measured tangentially on crown wheel teeth. Movement of ring nut one serration affects backlash by about .005in, but locking tab can be relocated to suit exact backlash.

CHASSIS

BRAKES

Girling hydraulic. Hydrastatic on early vehicles, with no adjustment. Handbrake operates Girling mechanical brake at rear of transfer box.

To adjust wheel brakes jack up wheel, turn snail cam adjuster until shoe binds, and back off until free.

To adjust handbrake turn square-ended adjuster until shoes make contact with drum, and back off two clicks. Apply brake firmly to centralize shoes. To reset linkage after overhaul adjust hand lever end of pull rod *after adjusting shoes* so that lever pulls up two notches on ratchet before applying brake.

BRAKE DATA		
	Front and Rear	Transmission
Drum diameter ...	10 in	9 in
Lining : length ...	8$\frac{13}{16}$ in	7$\frac{17}{32}$ in
width ...	1$\frac{1}{4}$ in	1$\frac{1}{4}$ in
thickness ...	$\frac{3}{16}$ in	$\frac{3}{16}$ in
No. of rivets per shoe...	10	10

SPRINGS

Semi-elliptic front and rear. Bonded rubber shackle and anchorage bushes. Plain bolts, tighten fully against inner member of rubber bush. Front springs are shackled at front, rear springs at rear. Front and rear bushes (sets of 6) not interchangeable. Up to vehicle No. 860647 all shackle and anchor bolts were same. Later anchor bolts not interchangeable. Up to vehicle No. 8665464, front spring shackles had thrust washers (.090, .100, .110 and .120in thick) on each side to limit side play to .005in.

SPRING DATA				
	Front		Rear	
	Up to Veh. No. 862114*	Veh. No. 862115 on	Near-side	Off-sdie
Length (eye centres, flat)	36$\frac{1}{4}$ in	36$\frac{1}{4}$ in	42 in	42 in
Width ...	1$\frac{3}{4}$ in	1$\frac{3}{4}$ in	1$\frac{3}{4}$ in	1$\frac{3}{4}$ in
No. of leaves	9	9	9	10
Free camber	3$\frac{1}{2}$ in	4 in	4 in	4$\frac{1}{2}$ in
Loaded camber	$\frac{3}{8}$ in	$\frac{3}{8}$ in	$\frac{3}{8}$ in	$\frac{3}{8}$ in
At load ...	617 lb	755 lb	660 lb	750 lb

* These dimensions are for service replacement spring. Previous springs had 8 leaves and 2$\frac{1}{2}$in or 3$\frac{1}{2}$in free camber

FRONT AXLE

Final drive assembly interchangeable with rear axle. Inner swivel housings flange-bolted to ends of axle casing enclose Tracta constant velocity universal joints. Driving members integral with half-shafts. Driven members integral with stub axle shafts, which are fully floating in hubs. Wheels run on taper roller bearings on stub axle tubes, which are flange-bolted to outer swivel housings.

Each inner swivel housing carries taper roller king pin bearings. Each half of two-piece king pin spigoted in outer swivel housing and registering

in inner race of bearing. Shims (.003, .005, .010, .020in thick) under shoulder of each king pin for bearing adjustment.

To dismantle axle disconnect brake fluid pipes, track rod and drag link, and undo six bolts on axle casing flange at each side. Hub, swivel housing and half-shaft assembly can then be drawn off. Take care not to damage oil seal inside end of axle casing.

Prise off hub cap, take off shaft nut and plain washer, and remove six set-screws holding driving member to hub. Scribe line on hub and driving member for correct refitting, and draw off driving member. Unscrew locknut and adjusting nut on stub axle tube, noting keyed tabwasher between and keyed thrust washer behind. Draw off hub with roller bearings. Oil seal at back will keep inner bearing together. Note distance-piece behind inner bearing, on which oil seal runs.

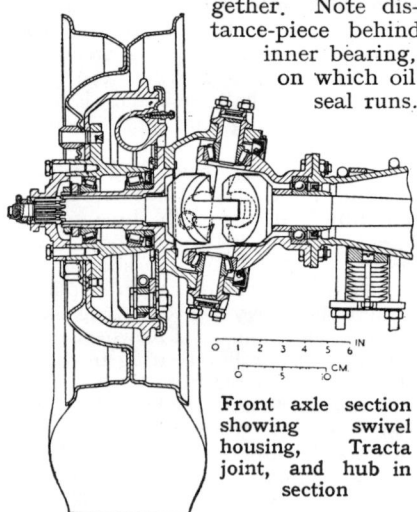

Front axle section showing swivel housing, Tracta joint, and hub in section

Take out six setscrews holding backplate and stub axle tube to outer swivel housing (when reassembling use Shakeproof washers). Remove brake backplate assembly and separate tube from outer swivel housing, drawing out stub axle shaft with male and female members of Tracta joint. Bush in inner end of stub axle tube should have .020-.028in clearance to shaft.

Detach oil seal retainer and seal from outer swivel housing (five short setscrews and one long lock stop set-screw and locknut). Detach steering lever and king pin assembly from top of swivel housing, and king pin from bottom, noting shims. Swivel housings can then be parted and rollers and inner races of bearings tipped out.

Half-shaft retained in ball bearing in inner swivel housing by collar pressed on. To remove half-shaft press out with at least 20 ton pressure. If not available, turn or cut collar off. When pressing on new collar see that distance-piece is fitted between bearing and shoulder on shaft.

When reassembling king pin bearings use shims to give .005in preload. Test end float of outer swivel housing on inner with dial gauge, and extract

shims until there is no float and no drag. Extract extra .005 shim to give preload, then equalize shims between upper and lower king pins.

When assembling Tracta joint fit female (slotted) member in claw end of half-shaft so that flat on running diameter is last part to engage claw. Part should slide easily into claw. Fit male member into claw end of stub axle shaft and engage with female member, holding stub shaft in place while stub axle tube is fitted to swivel housing. It is possible to fit male and female members of joint wrong way round, but force would be needed.

When assembling hub, adjust bearings to give .002-.004in end float.

Steering ball joints sealed side-plug type, pre-lubricated. Renew as assembly. Shanks threaded left- and right-hand, screwed into tubes and clamped. All six joints are identical except for thread of shank.

Adjustable lock stops have different settings for different tyres. Dimension from face of oil seal retainer to top of bolt head is $\frac{7}{16}$in for 6.00-16 tyres, $\frac{23}{32}$in for 7.00-16 tyres.

Steering relay lever and shaft assembly consists of tubular housing bolted vertically to cross-member, and carrying shaft in two split Tufnol conical bearings, which are heavily spring-loaded to damp steering.

To remove relay assembly detach grille, take out bolts holding grille panel to wings and chassis frame, and extract rubber washers under panel. Disconnect fore-and-aft drag link from upper lever, and draw off lower lever (cotter-clamped). Assembly can then be pushed out upwards after removal of two bolts.

To dismantle relay assembly detach upper lever, both end caps with oil seals, and brass thrust washers. Cover one end of shaft with heavy rag and tap shaft out, taking care as first split Tufnol bush is exposed, as spring is compressed to over 100lb. Release gently and tap shaft out with second bush. Keep pairs of bushes together. Spring data :—

No. of working coils		10
Free length ..	..	7$\frac{1}{4}$in
Fitted length	..	3in
At load ..	..	104$\frac{1}{2}$lb

To reassemble relay secure one split bush to shaft with 2in hose clip. Compress spring to 3in and fit special clips. Fit spring and second Tufnol bush on shaft, with washers at each end of spring, and clip second bush in place. Then carefully extract spring clips and push assembly into housing, freeing hose clips as bushes are pushed home. When assembly is complete and filled with oil, it should need at least 12lb on lever to turn shaft.

STEERING DATA		
Castor	...	3 deg.
Camber	...	1$\frac{1}{2}$ deg.
King pin inclination ...	...	7 deg.
Toe-in	...	$\frac{3}{32}$-$\frac{3}{16}$ in
No. of turns lock to lock ...	...	2$\frac{1}{4}$

SPECIAL TOOLS

Jig block for reboring, from Rover Co., part No. T 1287
Clip for hydraulic plunger on timing chain damper
Extractor for water pump impeller
Extractor for exhaust rocker shaft
Extractor for camshaft sprocket
Right-angle O/E spanner ($\frac{5}{16}$ in. B.S.F.) for carburettor flange nuts
Protector for transfer box oil seal
Packing pieces for transfer box output shaft bearings
Bar spanner for differential bearing locknuts
Clips for dial gauge attachment
Clips and bars for assembling steering relay unit

STEERING GEAR

Burman Douglas worm and nut.

To remove gear from vehicle remove battery, air cleaner, battery box and air cleaner support. Disconnect column wiring from junction box, extract control tube and draw off steering wheel (clamped on serrations). Release column from steady bracket and draw off drop arm (clamped on serrations). Detach gear from chassis frame and lift out over front.

Only adjustment is for column end

play, by thin nut and locknut concealed by wheel hub.

SHOCK ABSORBERS

Woodhead Monroe telescopic. Early front No. 453. Later both No. 454. No attention needed.

BODY

Truck body (aluminium panels, steel cappings) bolted to seat box under wheel arches and to rear cross-member. Instrument panel attached to dash by four screws. Detach for access to instrument wiring.

POWER TAKE-OFF & PULLEY

P.T.O. shaft at rear driven by propeller shaft from rear of gearbox through spur reduction gear in housing bolted to rear cross-member. Gears, 20 and 24 teeth, can be interchanged to give alternative ratios.

Pulley driven through spiral bevel gears in separate housing bolted to

GENERAL DATA

Wheelbase		6ft 8in
Track, front and rear		4ft 2in
Turning circle : 6.00—16 tyres ...		35ft 0in
7.00—16 tyres ...		40ft 0in
Ground clearance		8$\frac{1}{2}$in
Weight (dry)		22$\frac{1}{4}$ cwt.
Tyres : either		6.00—16
or		7.00—16
Overall length		11ft 0in
Overall width		5ft 1in
Overall height : with screen ...		5ft 5$\frac{1}{2}$in
with hood ...		5ft 10in

p.t.o. housing, and fitting over shaft splines.

To dismantle p.t.o. detach bearing caps and oil seal housings. Undo shaft nuts and tap shafts out of bearings and gears. Detach large bearing housings and remove gears. Both shafts run on taper roller bearings (all interchangeable). Outer races located in housing by spring rings. Inner races pulled up against gears by shaft nuts, with shims (.005, .010, .020in thick) for bearing adjustment, so that they are free without play. Note that bolts on propeller shaft flange are retained by spring ring.

To dismantle pulley assembly draw off pulley with flange, and separate pulley shaft and bearing assembly from driving shaft housing (flange-bolted with shims for mesh). Tap shaft out of taper roller bearings.

Inner races of bearings separated by distance-piece with shims (.005, .010, .020in thick) for bearing adjustment.

Detach driving shaft end cap (flange-bolted with shims for bearing adjustment) with outer race of rear taper roller bearing. Inner race pressed on to hub of bevel pinion, which is retained on splined shaft by setscrew and large washer with cork washer behind. Shims between pinion and shoulder on shaft for mesh. Adjust all bearings to be free without play. Note that seal at outer end of pulley shaft is fitted with lip outwards to exclude dirt.

LAND-ROVER WIRING DIAGRAM

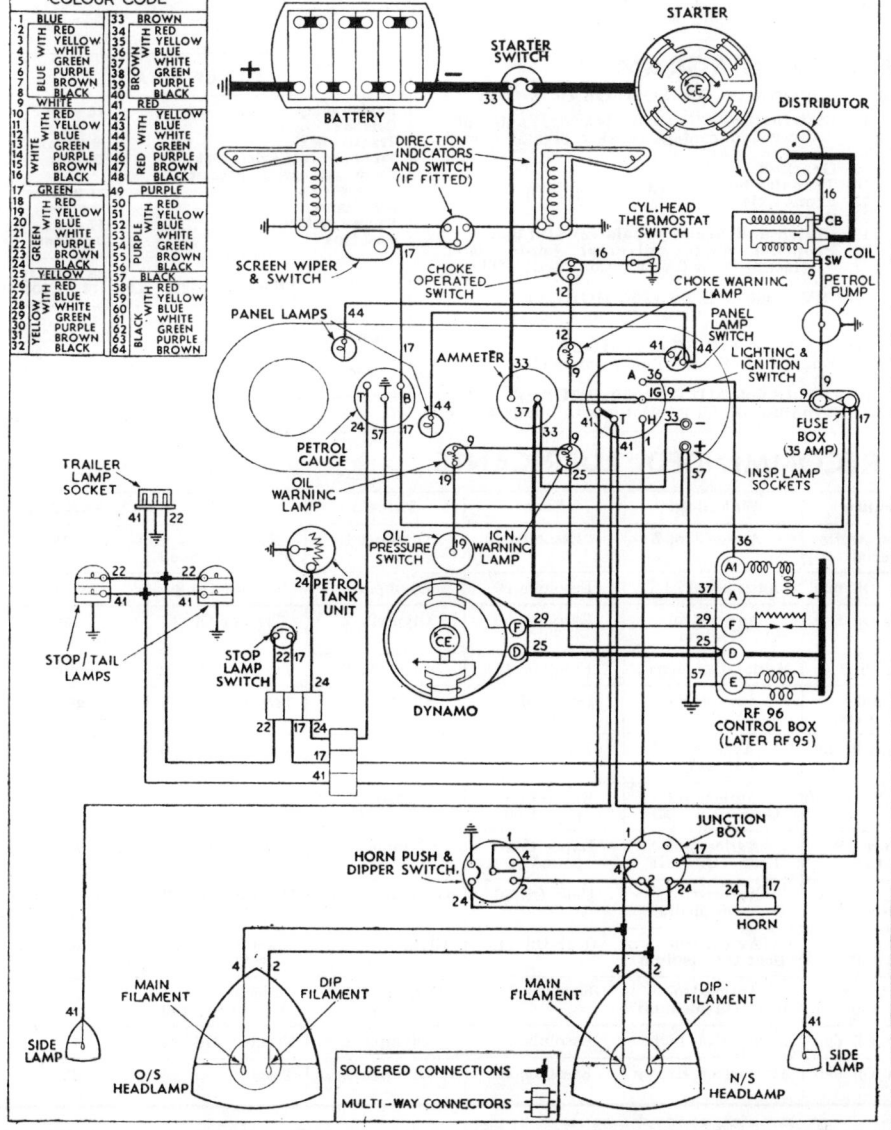

ELECTRICAL DATA
Lucas Equipment

	Model	Service No.
Dynamo : early	C39P	22255A
later	C39PV	22250F
Starter : early	M418G	255666
later	M418G	25514B
Starter switch	ST18	76401B
Lighting and ignition switch	PLC6	34035A
Control box : early ...	RF96/2	37098A
later ...	RF95	3706:E
Battery	FTW9A	—
Distributor	DVXH4A	40154A
Coil	B12	45012A
Fuse box (with RF96 control box)	SF5/2	385175
Headlamps	L/WD/HO	50124B
Sidelamps	451	52047A
Stop/tail lamps ...	ST51/1	53072B
Screenwiper	CW1	730292
Horn	HF1235	70036A

BULBS

	Voltage	Wattage	Lucas No.
Headlamps	12	36/36	171
Side, stop/tail, panel lamps	12	6	207
Warning lamps (M.E.S.)	2.5	.2	970

FUSES

Accessories	35 amperes	FA35

35

LAND-ROVER MAINTENANCE DIAGRAM

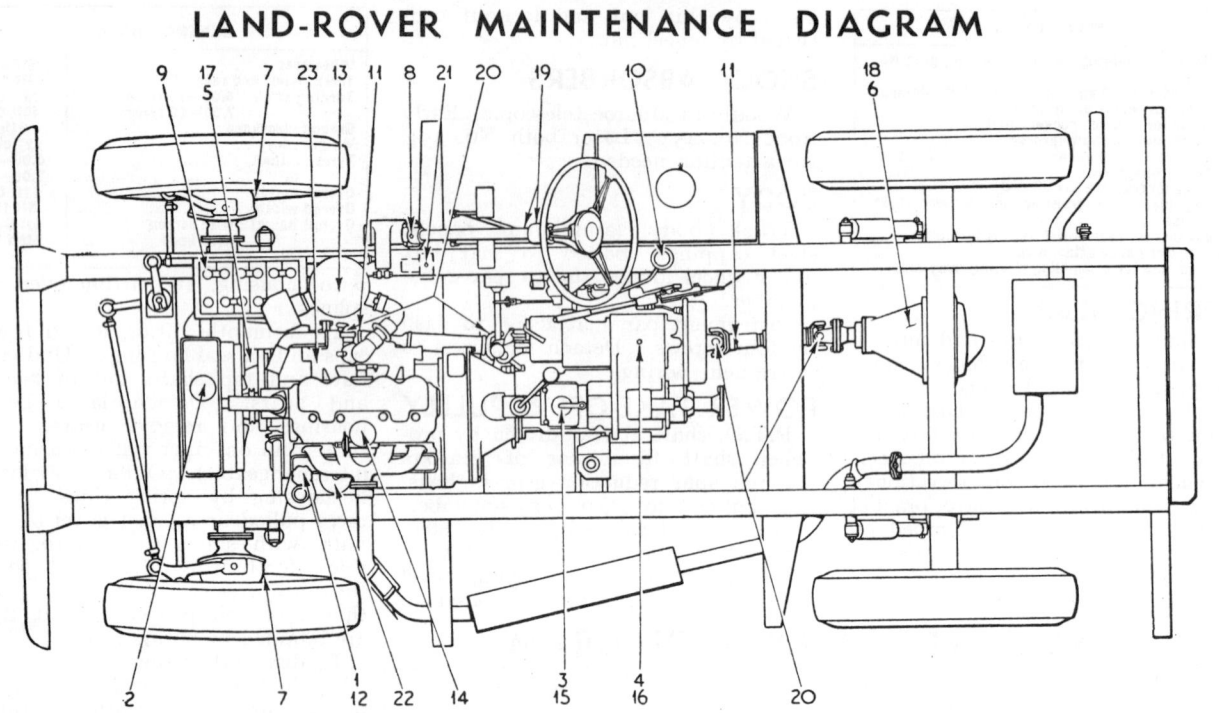

KEY TO MAINTENANCE DIAGRAM

PERIODS ARE QUOTED IN HOURS FOR STATIONARY OR LOW GEAR WORK

DAILY
1. Engine sump } Top up
2. Radiator

EVERY 1,000 MILES (30 HOURS)
3. Gearbox
4. Transfer box
5. Front axle
6. Rear axle
7. Swivel housings } Top up
8. Steering box
9. Battery
10. Brake fluid reservoir
11. Propeller shaft splines (2)—Oil gun

EVERY 3,000 MILES (100 HOURS)
12. Engine sump—Drain, clean intake filter and refill
13. Air cleaner—Drain, clean and refill bowl with engine oil. Swill filter gauze in petrol

14. Distributor—Oil shaft bearing, auto advance and contact breaker pivot. Grease cam
15. Gearbox
16. Transfer box } Drain and refill
17. Front axle
18. Rear axle
19. Clutch and brake pedals (2)—Oil gun
20. Propeller shaft universal joints (later vehicles) (4, or 6 with p.t.o. fitted)—Oil gun

EVERY 6,000 MILES (200 HOURS)
21. Petrol pump—Clean filter and sediment bowl

EVERY 9,000 MILES (300 HOURS)
22. Engine oil filter—Renew complete

EVERY 18,000 MILES (500 HOURS)
23. Dynamo—Refill lubricator with h.m.p. grease

FILL-UP DATA

Engine sump	10 pints
Gearbox	4 pints
Transfer box	6 pints
Front and rear axles (each)	3 pints
Front axle swivel housings (each)	1 pint
Air cleaner	2 pints
Cooling system	17 pints
Fuel tank	10 gallons

Tyre pressures :	Normal	Loaded	Cross-country 6.00—16	7.00—16
front	20 lb	20 lb	15 lb	13 lb
rear	26 lb	30 lb	20 lb	18 lb

RECOMMENDED LUBRICANTS

		Vacuum	Wakefield	Esso	Price's	Shell	SAE No.
Engine, Air Cleaner, Governor	Below 10° F	Mobiloil Arctic Special	Agricastrol Z	Essolube 10	Olympia 10	X.100, SAE 10 or Silver Shell	10 W
	10—32° F	Tractor Oil 620	Agricastrol LT	Essolube 20	Olympia F	Tractor Oil 20	20 W
	32—90° F	Tractor Oil 630	Agricastrol Medium	Essolube 30	Olympia M	Tractor Oil 30	30
	Over 90° F	Tractor Oil 640	Agricastrol Heavy	Essolube 40	Olympia Y	Tractor Oil 40	40
Gearbox, Transfer box	Below 10° F	Mobilube CW Special	Agricastrol Medium	Gear Oil 80	Olympia Y	Tractor Oil 30	80 gear or 30 engine
	Over 10° F	Tractor Oil 650	Agricastrol Heavy	Essolube 50	Olympia O	Tractor Oil 50	50
Differentials, Tracta joints	Below 10° F	Mobilube GX 80	Agricastrol Gear Oil EP	Expee Compound 80	Energol EP SAE 80	Spirax 80 EP	80 EP
	Over 10° F	Tractor EP Gear Oil	Agricastrol Gear Oil EP	Expee Compound 90	Olympia EP	EP Tractor Gear Oil	90 EP
Steering box, Oil gun nipples	Below 10° F	Mobilube CW Special	Agricastrol Gear Oil Medium	Gear Oil 80	Olympia Amber	Tractor Oil 30	80 gear or 30 engine
	Over 10° F	Tractor Gear Oil 140	Agricastrol Gear Oil Medium	Gear Oil 140	Olympia Gear D.K.	Tractor Gear Oil SAE 140	140
Steering relay lever (sealed)		Tractor Gear Oil 140	Agricastrol Gear Oil Medium	Gear Oil 140	Olympia Gear D.K.	Tractor Gear Oil SAE 140	140
Power take-off, Pulley		Tractor Oil 620	Agricastrol Light	Essolube 20	Olympia F	Tractor Oil 20	20 W
Capstan winch		Tractor Oil 640	Agricastrol Heavy	Essolube 40	Olympia Y	Tractor Oil 40	40

Modified Land-Rover and New Pick-up

Increased Payload Space and Improved Bodywork : Pick-up Has Wheelbase of 8 ft. 11 in. and Can Carry 10 Passengers

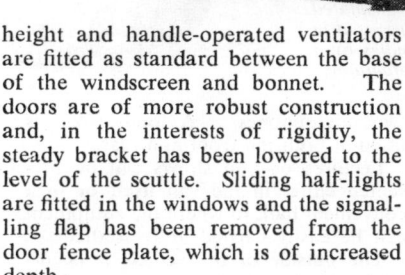

AN increase in the wheelbase from 80 in. to 86 in., a 25 per cent. greater payload space and a number of bodywork improvements are features of the new Land-Rover which was announced yesterday by The Rover Co., Ltd., Solihull, Birmingham. An important addition to the range is a pick-up truck with a wheelbase of 107 in.

The pick-up is similar to the hard-top version of the Land-Rover from

Cab improvements, common to both the Land-Rover and pick-up, include a facia panel with large instruments and two panel compartments, and pedals set at the conventional angle in a sloping toeboard.

the front bumper to the back of the cab, and, apart from the hood fittings and so on, the related modifications are also common to the standard Land-Rover.

The extended wheelbase of the Land-Rover, in conjunction with longer rear springs, improves the suspension, and the angularity of the propeller shaft is reduced. The length of the springs has been increased from 42 in. to 48 in., and the spring rate has been changed from 165 lb. per in. to 180 lb. per in.

The facia panel has large instruments of the car type, wide parcel compartments on each side and a neat grab rail formed by the top ledge. The pedals are mounted at a conventional angle in a sloping toeboard, and the gear lever is extended to within a short distance of the steering wheel.

The scuttle has been increased in

height and handle-operated ventilators are fitted as standard between the base of the windscreen and bonnet. The doors are of more robust construction and, in the interests of rigidity, the steady bracket has been lowered to the level of the scuttle. Sliding half-lights are fitted in the windows and the signalling flap has been removed from the door fence plate, which is of increased depth.

Large-section sealing rubber is fitted all round the doors, and on the standard Land-Rover the hood rail forms a rain gutter in which the canopy is a close fit. A channel-section gutter is attached to the forward hoop to act as a stay, the door closing against the stay to make a seal and to give support to the hood structure.

The raised scuttle enables the windscreen to be folded flat, with the spare wheel mounted on the bonnet. The screen is supported in this position by stays attached to the top rail, which are held by the bonnet catches. When not in use the stays are folded back along the rail.

Other modifications to the Land-Rover include removal of the spare-wheel well at the back to provide a flat floor, although the rear compartment may still be used for storing the wheel.

Body length behind the front seat has been increased by 9 in. and the unladen weight (with five gallons of petrol) by 154 lb. to bring the total up to 2,702 lb. The payload additional to the weight

The available space in the pick-up behind the wheel-arch allows useful lockers to be fitted for storage. The higher sides give better back support.

The pick-up has a wheelbase of 8 ft. 11 in. and can carry a payload of 1,120 lb. in addition to the driver and two passengers.

of two passengers remains unaltered at 1,000 lb.

The Rover four-cylindered engine developing 52 b.h.p. at 4,000 r.p.m. and the standard transmission components, with the same gear ratios, are fitted to all the vehicles in the range. Except for the extensions necessitated by the longer wheelbase, the chassis frames are similar to the former design.

Improved Cooling

The radiator has a pressure cap to raise the boiling point and obviate loss of water through surge. An oil-bath air cleaner is standard equipment and the oil filter is of the by-pass type. A pump intake filter is also fitted.

When the high-ratio transfer box is in operation, the selection of four-wheel drive is optional, but the drive is automatically engaged when the low-ratio transfer box is in use. The first-gear ratios are 16.17 to 1 and 40.69 to 1 respectively when the high-ratio range and low-ratio range of the transfer box are engaged. The corresponding top-gear ratios are 5.39 to 1 and 13.58 to 1.

The unladen weight of the pick-up truck is 3,088 lb., and the payload is 1,120 lb. The rear compartment extends 6 ft. behind the cab and a locker is fitted behind the wheel-arch on each side. The height of the body sides has been increased by about 6 in. compared with the height of the Land-Rover.

A de-luxe model having numerous cab refinements is also marketed. These include a head lining of plastic-covered felt, three cushions, a thick one-piece back rest covered with plastic-woven fabric, blue leathercloth lining of the instrument panel and of the parcel compartments, doors and rear panels, padded waist rails and the addition of door pockets.

A canvas hood, a tonneau cover and rear seats are available for the rear compartment. There is room for eight passengers in addition to the driver and two front-seat passengers.

Prices of Land-Rover models are: Basic (7 ft. 2 in. wheelbase), £570; pick-up (standard), £635; pick-up (de luxe), £655; fire engine, £905; welding outfit, £915.

A Miniature Mobile Film Unit for Haiti

A MOBILE advertising and film unit based on a Land-Rover chassis has been completed for J. Bibby and Sons, Ltd., soap manufacturers, by J. H. Jennings and Son, Ltd., Sandbach, Cheshire. It is intended for use in Haiti, West Indies.

Constructed throughout in light alloy, the body has a layer of insulating material sandwiched between the interior and exterior panels to enable the interior to be kept cool.

The interior is entered at the rear, where there are a tailboard and a large top-hinged canopy. Cupboards are incorporated along both sides, and above them is a rack on which two projection screens are carried. Additional storage space is provided in a Luton compartment.

This Land-Rover with Jennings insulated body is being used in the West Indies by a soap manufacturer.

A detachable bench is provided for rewinding or splicing film, and the equipment includes a B.T.H. 16 mm. sound film projector, telescopic projector stand, rewinder, splicer, tape recorder, record player, three fans and two microphones. A loudspeaker is fitted at both the front and rear.

Lectures or film shows can be presented either in the van itself or in a convenient hall, to where the equipment can be easily carried and-installed.

The outside of the vehicle is decorated with reproductions of some of the operator's products.

This is one of several specialized vehicles which J. H. Jennings and Son, Ltd., have recently completed. Two mobile lending libraries have also been built for West Riding County Council. They are based on Bedford 4-ton long-wheelbase chassis converted to full forward control and with frames extended by Baico.

Mobile Dispensary for the Desert

A FOUR-WHEEL-DRIVE long-wheelbase Land Rover forms the basis of a mobile dispensary which the British Red Cross Society are sending to Dubai for operation in remote areas of the Persian Gulf. It has been built and equipped by Messrs. Pilchers, ambulance specialists, 314 Kingston Road, London, S.W.20.

The bodywork, which incorporates an aluminium double-skin roof insulated with Isoflex, is well ventilated. Cupboards for drugs, dressings and other equipment are fitted along the off side, whilst on the near side there is a light alloy folding stretcher. Fresh water is carried in Polythene containers, and washing facilities are provided. Hooks are provided along the sides from which water skins can be suspended.

Additional accommodation is afforded by a tent fixed to the rear of the body. The tailboard forms a folding step.

Special mattress devices are provided for preventing the wheels from sinking in the sand when the unit is operating in the desert.

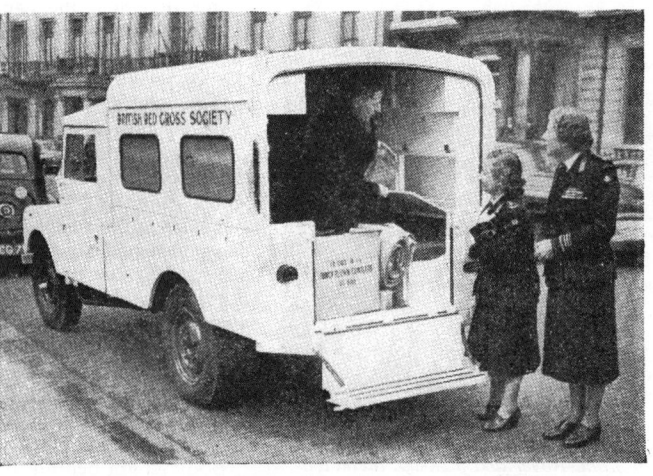

Mr. R. Pilcher explaining features of the dispensary to officials of the British Red Cross Society before it left this country for Dubai.

With a Land-Rover in Argentina

NEWS comes from R. K. Hubbard who, many years ago, was secretary of the Mechanical Transport Committee at the War Office. During his 16 years' military service and even before 1913, when the late George Watson was technical editor, he was a reader of *The Commercial Motor*. When he went to Argentina in 1930 to become an executive of the British Argentine Railways, he continued to receive it, but he retired in 1948, as the railways were taken over by the government of that country.

He remained there and is now the proud possessor of a Land-Rover which, he says, was the only British-built vehicle he could buy new and it has turned out most suitable for the long journeys, often over unmade roads, which he makes. He has fitted out this go-anywhere model for, as he says, " sleeping, eating and drinking."

His principal journeys are from Buenos Aires to the Parana River, where he goes dorado fishing, this species being as good as salmon. The mileage is about 700 each way. When Buenos Aires gets too hot in December or January, he travels to the Andes —a thousand miles away—for trout and salmon fishing. Most of the journey is over rough gravel roads, with stretches of soft sand and mudholes in different parts, and he usually limits himself to 300 miles a day.

He states that he observed with interest the founding of the Institute of Road Transport Engineers, which he considers was an excellent move.

In April, Mr. Hubbard is returning to London for a few months and hopes to meet some of his old friends.

NOTHING could stop the Land-Rover in over 250 miles of road and cross-country trials in which it carried a full load, and hauled a trailer with 15-cwt. load up the notorious Succombs Hill, in Surrey, with its 1 in 4¼ gradient. Lengthening the wheelbase of the Land Rover has not impaired its willingness for hard work, whilst the degree of riding comfort can now be classified as superior, and there is considerably more body space.

This was especially noticeable in the 107-in. wheelbase model tested, which was supplied with de luxe trim including an upholstered bench-type seat, lined roof, padded waistline and arm rests, and leathercloth covering to all metal parts in the cab. Additional pampering for the driver is represented in the heater and demister unit

(*Above*) *Carrying a full payload, the Land-Rover showed its ability to climb steep loose-surfaced gradients in the cross-country trials.* (*Left*) *Large-section cross-country tyres were borrowed from an operator for off-the-road tests, the four wheels complete being changed in under five minutes.*

Latest Land-Rover has More Payload Space: Detail Mechanical Modifications have Provided a Higher Standard of Cross-country Performance and Improved Braking

by Laurence J. Cotton, M.I.R.T.E.

Smooth and Willing— the Big Land-Rover

which was particularly welcome in the cold morning air.

In general, the Land-Rover can be supplied in standard form with enclosed driver's cab and individual seats for the driver and two passengers, or it can be equipped with car refinements for the gentleman-farmer, who might require the vehicle for social purposes as well as for use on the land.

It has the 2-litre, 52 b.h.p. engine with overhead-inlet and side-exhaust valves, unit mounted with a four-speed constant-mesh gearbox in which third and top ratios have synchromesh engagement. The drive is passed through a two-speed transfer box which also contains the gears for selective engagement of four-wheel drive. Formerly the Land-Rover had permanent drive to the front wheels with a free-wheel to prevent transmission wind-up on hard-surfaced roads. Dispensing with this enables the vehicle to be driven as a normal rear-wheel-drive machine on the road when high transfer gear is engaged. Four-wheel drive can be brought into use by depressing a knob control. It is automatically engaged when the low transfer ratio is in use.

Crossing a ditch at an angle showed the remarkable stability of the vehicle. The eight forward ratios ensure that there is always a gear suited to any kind of work.

The Land-Rover finds a soft spot with the near-side wheels, but with four-wheel drive engaged, it was extricated without assistance. Nothing baulked the vehicle during the cross-country trials.

Brockhouse 15-cwt. all-steel trailer which also carried a full payload. The Brockhouse trailer has a 16-gauge floor and 18-gauge side and end panels, and is based on a sturdy rolled-steel channel-section chassis. It has a body measuring 6 ft. by 3 ft. 2 in. and the sides are 1 ft. 6 in. deep.

At the Brompton railway sidings, which was the first stop after leaving the Rover service depot in London, the Land-Rover tipped the scales at 2 tons, complete with its load, and the Brockhouse trailer, with 15 cwt. of sand, added another ton to the gross weight. The trailer weighed 5 cwt. unladen.

When driving in London traffic I found normal low gear was required for smooth acceleration from rest when pulling the loaded trailer, but second gear could be used comfortably when operating solo. Four-wheel drive was required on one occasion following wheelspin when moving off on a slippery-surfaced gradient and pulling the loaded trailer. The machine was equipped with standard-tread tyres for the road trials.

Although driven hard in traffic and over country roads, the vehicle behaved well and showed no indication to wander at the back when cornering on a heavily cambered road. The braking was impaired when hauling the trailer, but the driver is usually more cautious under such conditions and allowance is made for an increased stopping distance.

New Braking System

It is understood that the current production Land-Rover chassis is now being equipped with a new braking system, using larger and wider shoes and facings, to enable the payload to be increased from 10 cwt. to 15 cwt. The new brakes are Girling hydraulic two-leading-shoe type at the front and leading and trailing shoe at the rear. They are 11 in. diameter and $2\frac{1}{4}$ in. wide.

The first run on Succombs Hill was made from a moving start, and the 1 in 5 and 1 in $4\frac{1}{2}$ gradients were climbed using the normal-low ratio. After this, stop-start tests were staged, and here the lower ratio of the auxiliary box was required.

The ample power available was demonstrated at this stage by using the second and third ratios of the main gearbox and releasing the accelerator almost back to the idling position after moving away from rest. With the trailer detached, the

With the larger engine and eight forward gears providing an overall spread, in conjunction with the final drive, of 5.396 to 40.676 to 1, the machine is virtually potted power, as I found during my trials when it was hauling a trailer. As evidence of this it is used by one municipality for towing disabled trolleybuses.

The 107-in. wheelbase chassis is provided with a standard body of just over 6 ft. internal length which, carrying a nominal 10-cwt. load, gives practically equal weight distribution on the axles. This constitutes an important factor in cross-country driving over loose or soft ground.

There are many body styles available on the Land-Rover basic chassis structure, the one supplied for test having a canvas tilt and inward-facing seats along the sides, the latter forming special equipment. The addition of the crew seats in the body increased the normal unladen weight to 1 ton $8\frac{3}{4}$ cwt., with a full fuel tank, but the model supplied was not

Load on Land-Rover	Load on trailer	m.p.g.	Average speed
$11\frac{1}{2}$ cwt.	15 cwt.	17.4	33.1
$11\frac{1}{2}$ cwt.	8 cwt.	18.05	33.1
$11\frac{1}{2}$ cwt.	No trailer	19.85	33.9
6 cwt.	No trailer	20.2	33.0

Fuel consumption rate under differing conditions of operation.

equipped with power take-off drive, which can be arranged at the centre or rear of the chassis.

In addition to an $11\frac{1}{2}$-cwt. load of sand the Land-Rover hauled a

Although the gross weight, when hauling the Brockhouse trailer, was 3 tons, the 1-in-4¼ gradient of Succombs Hill was climbed and stop-start tests made with the aid of the low transfer ratios.

Land-Rover romped up the hill in second gear from a moving start and normal-low ratio sufficed after stopping on the steeper sections.

The trailer was hitched up for the initial consumption test, the combined payload being 1 ton 6½ cwt. This trial was conducted over a 20-mile out-and-return course between Godstone and East Grinstead and included normal main-road undulations of up to 1 in 16. An auxiliary petrol tank was fitted and the speed controlled to 35-38 m.p.h. Under such conditions, a fuel return of 17.4 m.p.g. at 33.1 m.p.h. was obtained.

Payload Reduced

I then removed 8 cwt. from the trailer which reduced the overall payload to 18½ cwt. This made relatively little difference to the performance on the gradients and gear changes were made at approximately the same points on the journey. There was, however, a small saving in fuel, the consumption rate being 18.05 m.p.g. at 33.1 m.p.h.

There was a marked difference with the trailer detached, and I noticed the improvement, with 13 cwt. less on the tail as represented by the trailer with part load. The Land-Rover was sharp off the mark and completed the course without need for indirect ratios apart from turning at the end of the outward run. Carrying an 11½-cwt. payload the return was 19.85 m.p.g. at 33.9 m.p.h. average speed.

Little Change in Consumption

To complete the trials I reduced the load by 5½ cwt. which would be generally representative of normal operation. This produced little effect from the economy angle, the consumption rate working out to 20.2 m.p.g. It appears likely that solo operation would generally give about 20 m.p.g., and 18 m.p.g. when used with a trailer.

Braking tests with full load produced an average stopping distance of 39 ft., and with the laden trailer attached this was increased to 45 ft., indicating well above average deceleration in both trials.

The cross-country tests were made without the trailer, and on general estate and agricultural work the Land-Rover did well on the standard road tyres fitted. Wheelspin baulked all efforts when attempting to negotiate ditches or loose-surfaced banks, so I borrowed a set of 7.50-16-in. Trakgrip tyres and wheels from an operator for the more arduous work. The Rover company recommend a

7.00-in. tyre as the largest size which should be fitted, but after further tests, when the suspension was fully tried out, I found no evidence of the covers having fouled the chassis, or noticeable difference in the steering.

The Land-Rover was put through strenuous exercises and made to climb seemingly impossible gradients. It failed on one, a loose-surfaced

bank of 1 in. 2½ incline, when the rear wheels churned through the soft soil and then failed to find further hold. When tackling this same gradient at speed, the tow hitch fouled the road, so I reversed and set an oblique course up the slope.

Had the load been less stable the truck might have overturned, such was the steepness of the slope, but

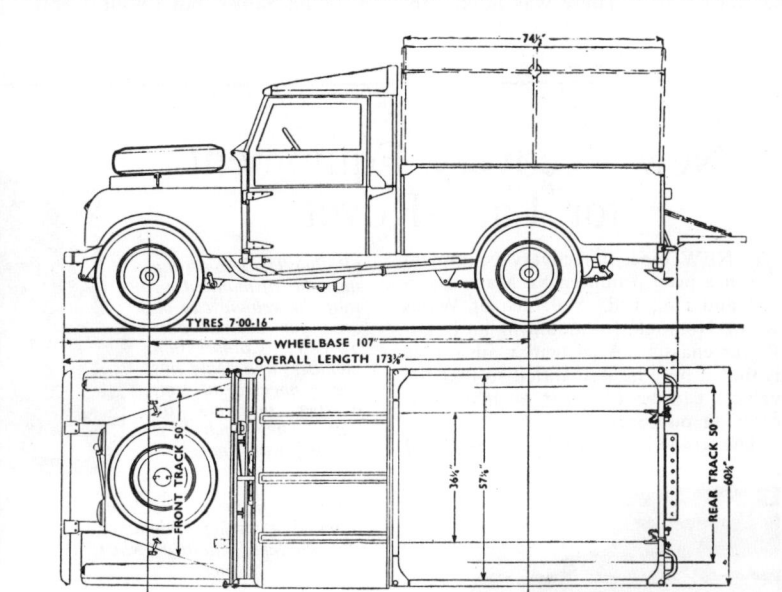

ROAD TEST No. 518—LAND-ROVER 107-in. WHEELBASE MODEL

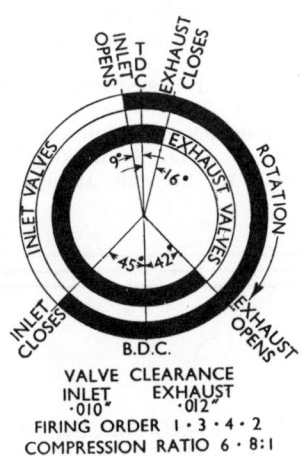

VALVE CLEARANCE
INLET   EXHAUST
·010   ·012
FIRING ORDER 1·3·4·2
COMPRESSION RATIO 6·8:1

MODEL: Land-Rover 107-in. wheelbase model with Brockhouse 15-cwt. trailer.

WEIGHTS:

	Tons	cwt.	qr.
Unladen (full tank) solo ..	1	8	3
Payload		11	2
Driver, observer, etc. ..		3	2
	2	3	3

DISTRIBUTION

Front axle ..	..	..	1	1	2
Rear axle ..	..	..	1	2	1

ENGINE: Rover four-cylindered overhead-valve petrol engine; bore 77.8 mm. (3.06 in.); stroke 105 mm. (4.134 in.); piston-swept volume 1.997 litres (121.9 cu. in.); maximum output 52 b.h.p. at 4,000 r.p.m.; R.A.C. rating 15 h.p.; maximum torque 101 lb.-ft. at 1,500 r.p.m.

TRANSMISSION : Through .9-in. diameter single-dry-plate clutch to four-speed gearbox and two-speed auxiliary box, thence by open propeller shafts to both axles.

GEAR RATIOS: 2.996, 2.043, 1.377 and 1 to 1 forward; reverse 2.547 to 1; rear axle ratio 4.7 to 1, auxiliary box ratios 2.888 and 1.148 to 1, giving overall gearbox ratios of 40.688, 27.742, 18.707, 13.578, 16.171, 11.026, 7.435 and 5.396 to 1.

BRAKES: Girling hydraulic system to leading and trailing shoes at all wheels. Hand brake operates on transfer box output shaft. Diameter of drums 10 in.; width of facings, 1½ in., total frictional area 100.8 sq. in., that is, 46.4 sq. in., per ton gross weight as tested.

FRAME: Welded box section.

STEERING: Burman worm and nut.

SUSPENSION: Semi-elliptic springs with Woodhead - Monroe telescopic shock absorbers.

ELECTRICAL : 12 v. compensated-voltage-control system with 51 amp.-hr. battery.

FUEL CONSUMPTION: With full load and no trailer 19.85 m.p.g. at 33.9 m.p.h. average speed, that is 43.5 gross ton m.p.g. as tested, giving a time-load-mileage factor of 1.473.

TANK CAPACITY: 10 gallons, range approximately 180 miles.

ACCELERATION: Through gears, 0–30 m p.h., 9.4 sec.; 0–40 m.p.h., 20 sec.

BRAKING: From 20 m.p.h., 17½ ft. (24.8 ft. per sec. per sec.); from 30 m.p.h., 39 ft. (24.8 ft. per sec. per sec.).

WEIGHT RATIOS: 1.19 b.h.p. per cwt. gross weight as tested.

TURNING CIRCLES: 50 ft. both locks.

MAKERS: The Rover Co., Ltd., Meteor Works, Birmingham.

this tactic was successful in overcoming the obstacle. I stopped the engine and paused for any sign of oil or petrol leakage, but there was neither, and the engine restarted without any marked trait of over-enriched fuel supply.

Although carrying an 11½-cwt. load the performance was beyond reproach and I found the Land-Rover capable of overcoming any obstacle where the tyres could find a reasonable grip. There was adequate

power, as proved by wheelspin on the 1 in 2½ slope, and the suspension provided smooth riding on the broken or corrugated surfaces.

The traffic indicators, fitted near the top of the windscreen, are in a vulnerable position when blazing a trail through overhanging trees, but otherwise the body and chassis are well protected in cross-country work. There appeared sufficient belly clearance when negotiating sharp-crowned banks, but I would prefer to

dispense with the spare wheel on the bonnet for improved near visibility in difficult cross-country work.

The performance of the vehicle exceeded my expectations under all conditions, and I have high regard for its economy, eight-speed gearbox, and ability to win through over nearly impossible ground. In de luxe form it has all the refinements which could be desired by an owner-driver, and long periods can be spent behind the wheel without strain.

New Recovery Equipment for Land-Rover

A NEW type of recovery equipment has been produced by Mann Egerton and Co., Ltd., 5 Prince of Wales Road, Norwich, for mounting on Land-Rover chassis. A particular advantage is that it may be dismantled so that the vehicle can be used for normal load-carrying purposes.

The crane is of 2-ton capacity and

(Right) The 2-ton crane may be dismantled so that the vehicle can be used for normal purposes. Compression members and cable ties are connected to supporting brackets by pins. The winch is a separate unit.

(Left) A frame at the front supports a towing ambulance and confers stability when towing.

the support brackets are welded to the chassis. The tubular compression members and cable ties are connected to the brackets by pins. The winch is built as a separate unit on a frame welded between the chassis members and situated immediately behind the bulkhead.

Chassis supports are provided for use when the crane is working to full capacity to relieve the springs of high

loading. The towing bracket is the standard M.E. type A. A bar, through which the hook passes, holds the load steady and at a constant distance from the rear.

A frame is welded between the front chassis members for supporting a towing ambulance. The ambulance housing confers stability to the steering when the vehicle is towing. Price of the equipment is £98.

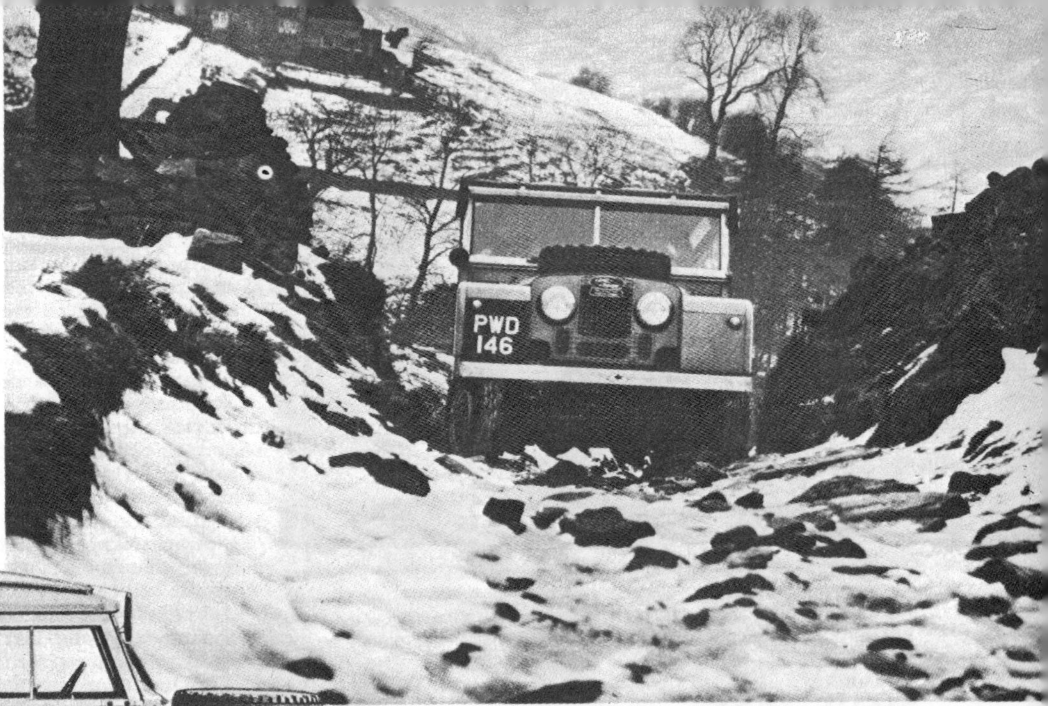

With four-wheel drive engaged, steep gradients with slippery surfaces could be climbed without the use of chains or special tyres

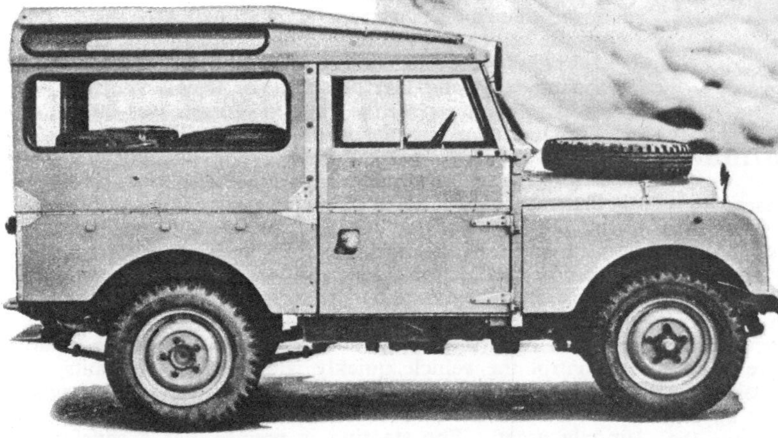

Built for a purpose as a working car, the Land-Rover Station Wagon makes no concessions to modern styling. Metal parts are galvanized. It is not good looking—nor is a tank—but both are admirably suited to their jobs

The *Autocar* ROAD TESTS

No. 1557: LAND-ROVER STATION WAGON

WEATHER conditions were hopeless for normal performance testing when George Mackie, of the Rover company, telephoned to suggest a trial of a Land-Rover Station Wagon in the snow. First-hand experience with this type of vehicle is always worth while, and although when the car was available there were signs of a noticeable rise in air temperature, which was followed by a quick thaw, the opportunity arose to try out the car on a variety of surfaces, including glazed roads, loose snow and very rough tracks.

The Station Wagon is an all-purpose vehicle based on the well-known Land-Rover, and containing the standard transmission with dual range gear box and optional four-wheel drive. In place of the familiar canvas top often fitted to the Land-Rover, there is a metal body, with four tip-up seats in the rear compartment, in addition to three-abreast seating in the front. In the design of the Station Wagon, emphasis has been laid on using as many of the standard Land-Rover components as possible. The result is a car that is essentially a Land-Rover, with additional equipment to meet the requirements of station wagon work. The robust character of the Land-Rover has been retained, and the familiar galvanized finish is used for usually bright parts such as windscreen frame and bumper.

A two-litre four-cylinder Rover engine is used, but for the Land-Rover application, fuel is metered by a down-draught Solex carburettor in place of the S.U. instrument used for Rover cars. The engine develops 52 b.h.p., and provides ample power for propelling a vehicle with a very large frontal area (by private car standards) at speeds approaching 60 m.p.h. The mean maximum speed recorded during the Road Test was 57.75 m.p.h., and although this is low compared with that of a two-litre saloon car, it is very creditable for a vehicle of this type. In spite of a modest maximum, on several occasions journeys of around 80 miles were covered without difficulty at an average speed of 40 m.p.h.

The eight-speed transmission is best considered in three stages. For normal road work it can be regarded as a conventional rear wheel drive, four-speed arrangement with synchromesh on top and third gears only. This layout is

To improve the heat insulation when the car is used in hot countries, the roof has a double skin, the outer surface of which is painted a light colour, and there is an air space between the two panels. There are additional windows in the lower roof panel, which considerably improve the rear passenger vision

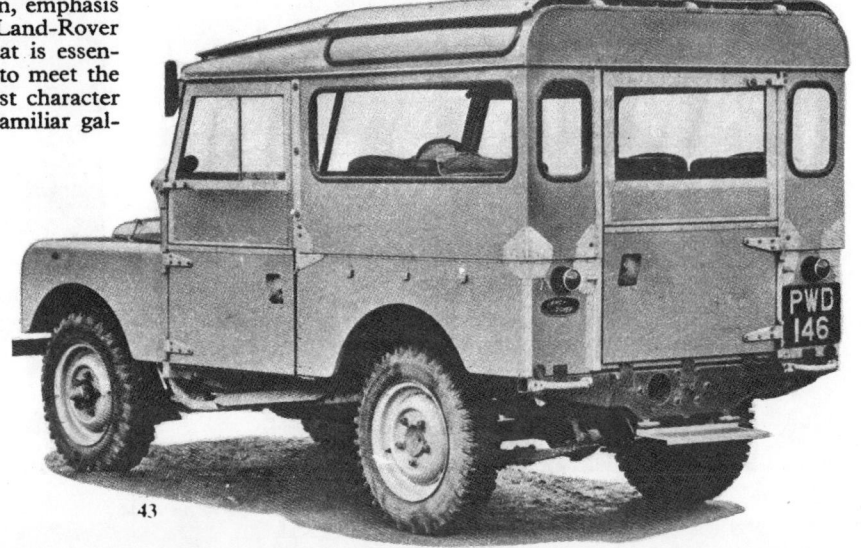

The floor in the rear compartment is covered with rubber, and four seats are provided, the cushions of which can be folded up when a clear floor area is required for carrying goods

ROAD TEST . . . . continued

very satisfactory, and the gears provide well-chosen steps with normal change-up points at 16, 27 and 40 m.p.h. Under adverse conditions, or when extra traction is required, four-wheel drive may be brought into operation by operating a small lever with a yellow knob.

During the performance testing, standing-start acceleration figures were taken both with two- and four-wheel drives engaged, and on a dry test surface the times were almost identical. The four-wheel drive, as well as increasing the traction when traversing rough colonial sections, is advantageous on slippery roads when, with all wheels pulling, the car can be accelerated quickly and in a straight line; in these conditions it will slide if similar tactics are attempted with only the rear wheels driving.

The third stage of the transmission—the transfer box—is operated by a three-position lever. This is kept in the forward position for all normal driving, but if it is moved to the rear position the overall top gear ratio is changed from 5.396 to 1 to 13.578 to 1, with corresponding torque multiplication for all the indirect ratios, including reverse. Movement of this transfer lever from high to low range automatically engages four-wheel drive, and disengages it again when the lever is moved to the normal drive range position. The third position on the transfer lever is neutral, which is used in conjunction with a power take-off.

To disengage the drive to the front wheels with normal drive range in use, it is necessary to move the transfer lever back to the low range position, and forward again into the normal driving position; this enables the yellow-handled lever to be disengaged. Because the front and rear wheels are normally rotating at the same speed, the four-wheel drive engagement can be affected when the car is in motion, but a down change on the transfer box should be made only when the vehicle is stationary. This arrangement of gearing and drive enables the vehicle to be driven almost anywhere and climb anything that provides the necessary grip.

As part of the test, the Station Wagon was driven up Hollinsclough, a motor cycle trials section near Buxton. The track was covered in snow, and in places there were rock steps, up to ten inches or a foot in height. The sides of the track were sloping in parts to form a V section gully. With four-wheel drive engaged, no difficulty was found in ascending the slope, and in fact several stops and restarts were made. Using ordinary road tyres, the increased wheel loading brought about by the weight of three or four passengers improved wheel adhesion.

When traversing rough country, it is necessary to be able to control the vehicle quickly within very fine limits, and consequently there are only 2¾ turns of the steering wheel from lock to lock; nevertheless, the effort required is not unduly great. The steering is precise, has a satisfactory self-centring action, and on the test car there was no noticeable lost motion. By private car standards the turning circle appeared to be larger than usual, but it must be remembered

The front compartment contains three seats. The fuel tank is placed below the driver's seat cushion; access is gained by removing the cushion and opening a hinged metal cover. A tool compartment is provided on the left-hand side of the car, and all the front seat squabs can be hinged down if required

that the use of four-wheel drive complicates the design, as the necessary angular steering movement must be provided in the front wheel drive transmission joints.

Like the open Land-Rover, the Station Wagon has non-independent front suspension, with half-elliptic leaf springs at both front and rear. The suspension is firm, and there is very little roll on corners. For normal road driving all the passengers have a comfortable ride, although it is a little harder than that usually found on normal private cars. Off the beaten track the suspension functioned well, providing sufficient flexibility to enable the wheels to follow uneven surfaces. No bottoming was noticed.

Hydraulically operated leading and trailing shoe brakes are used for both the front and rear wheels, and under test conditions these returned very good figures. No brake fade was experienced, and at all times the brake pedal had a firm and solid feel. In place of the usual arrangement by which the handbrake is mechanically connected to the rear brakes, a transmission handbrake is provided on the Station Wagon, and this proved adequate for parking purposes.

To provide suitable ground clearance the Land-Rover is necessarily a high vehicle, so that the driver has a commanding position which enables him to see over the top of hedges instead of half-way up them. All-round visibility is also very good, although forward vision would be improved if the spare wheel were not carried on the bonnet top. The rear view, however, is barely adequate, the right-hand wing mirror providing the driver with very limited rear vision. This would be much improved if an additional internal mirror were provided, and this could be conveniently mounted on the column which divides the two-piece windscreen.

Although the driving seat is non-adjustable, it is well positioned and very comfortable, and the general layout of

LAND-ROVER STATION WAGGON

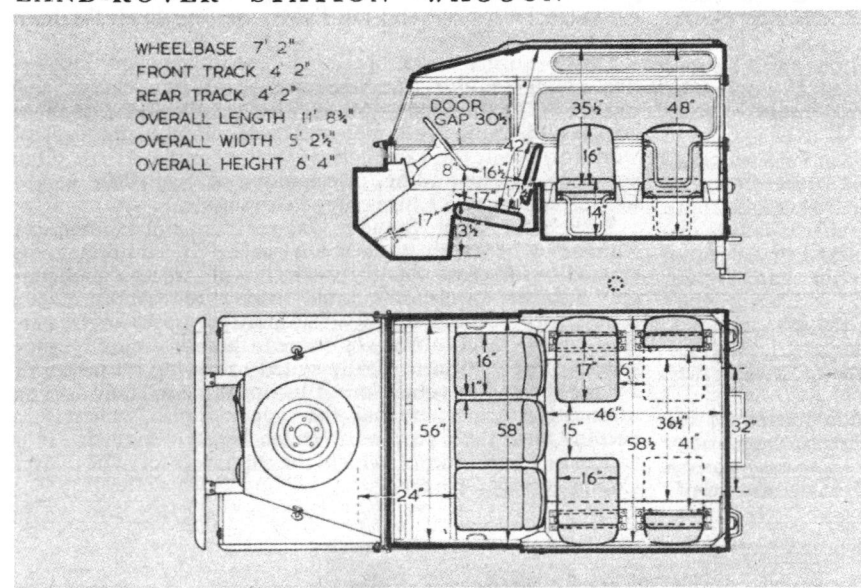

WHEELBASE 7' 2"
FRONT TRACK 4' 2"
REAR TRACK 4' 2"
OVERALL LENGTH 11' 8¾"
OVERALL WIDTH 5' 2½"
OVERALL HEIGHT 6' 4"

Measurements in these ⅛in to 1ft scale body diagrams are taken with the driving seat in the central position of fore and aft adjustment and with the seat cushions uncompressed

──── DATA ────

PRICE (basic), with station wagon body, £630 0s 0d.
British purchase tax, £263 12s 6d.
Total (in Great Britain), £893 12s 6d.
Extras: Radio, £28 19s 6d; Heater, £14 19s 0d.
ENGINE: Capacity: 1,997 c.c. (121.8 cu in).
Number of cylinders: 4.
Bore and stroke: 77.8 × 105 mm (3.063 × 4.134 in).
Valve gear: overhead inlet, side valve exhaust.
Compression ratio: 6.7 to 1.
B.H.P.: 52 at 4,000 r.p.m. (B.H.P. per ton laden 34.6).
Torque: 101 lb ft at 1,500 r.p.m.
M.P.H. per 1,000 r.p.m. on top gear, 15. (6 in Low Range).
WEIGHT: (with 5 gals fuel), 26½ cwt (2,968 lb).
Weight distribution (per cent): F, 52.7; R, 47.3.
Laden as tested: 30⅛ cwt (3,368 lb).
Lb per c.c. (laden): 1.69.
BRAKES: Type: F, leading and trailing; R, leading and trailing.
Method of operation: F, hydraulic, R, hydraulic.
Drum dimensions: F, 10in diameter; 1½in wide. R, 10in diameter; 1½in wide.
Lining area: F, 52.35 sq in. R, 52.35 sq in. (69.5 sq in per ton laden).
TYRES: 6.00—16in.
Pressures (lb. per sq in): F, 25; R, 25 (normal). F, 15; R, 15 (for exceptionally soft ground).
TANK CAPACITY: 10 Imperial gallons.
Oil sump, 10 pints.
Cooling system, 17 pints (plus 1 pint if heater is fitted).
TURNING CIRCLE: 37ft 0in (L and R).
Steering wheel turns (lock to lock): 2¾.
DIMENSIONS: Wheelbase: 7ft 2in.
Track: F, 4ft 2in; R, 4ft 2in.
Length (overall): 11ft 8¾in.
Height: 6ft 4in.
Width: 5ft 2½in.
Ground clearance: 8in.
Frontal area: 28 sq ft (approximately).
ELECTRICAL SYSTEM: 12-volt; 51 ampère-hour battery.
Head lights: Double dip; 42—36 watt bulbs.
SUSPENSION: Front, half-elliptic leaf springs. Rear, half-elliptic leaf springs.

──── PERFORMANCE ────

ACCELERATION: from constant speeds. Speed Range, Gear Ratios and Time in sec.

M.P.H.	5.396 to 1	7.435 to 1	11.026 to 1	16.171 to 1
10—30..	11.4	8.1	6.9	—
20—40..	12.6	10.8	—	—
30—50..	18.2	—	—	—

From rest through gears to:

M.P.H.	Rear wheel drive sec.	Four wheel drive sec.
30	7.1	7.0
50	25.5	24.9

Standing quarter mile, 25.7 sec, with four wheel drive, 25.4 sec.

SPEED ON GEARS:

Gear	M.P.H. (normal and max.)	K.P.H. (normal and max.)
Top (mean)	57.75	92.94
(best) ..	59.5	95.76
3rd	40—46	64—74
2nd	27—32	43—51
1st	16—22	26—35

TRACTIVE RESISTANCE: 47 lb per ton at 10 M.P.H.

SPEEDOMETER CORRECTION: M.P.H.

Car speedometer:	10	20	30	40	50	60	64
True speed:	9	18	27.5	36.5	45	55	59.5

TRACTIVE EFFORT:

	Pull (lb per ton)	Equivalent Gradient
Top	200	1 in 11.1
Third	270	1 in 8.2
Second..	370	1 in 6.0

BRAKES:

Efficiency	Pedal Pressure (lb)
92 per cent	125
78 per cent	100
63 per cent	75
48 per cent	50

FUEL CONSUMPTION:
21 m.p.g. overall for 315 miles (13.45 litres per 100 km).
Approximate normal range 19—26 m.p.g. (14.9—10.9 litres per 100 km).
Fuel, First grade.

WEATHER: Fine, dry surface, moderate wind across course.
Air temperature 38 deg. F.
Acceleration figures are the means of several runs in opposite directions.
Tractive effort and resistance obtained by Tapley meter.
Land-Rover described in *The Autocar* of April 30, 1948.

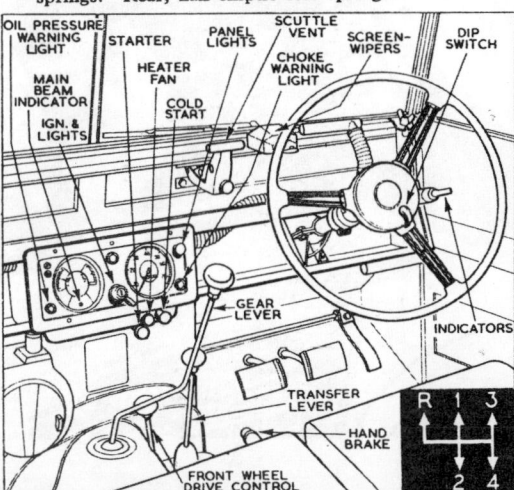

seat and pedals is such that a large range of seat adjustment is unnecessary. The pedals are well spaced and comfortable to operate, and side flanges are provided on the clutch and brake pedals to reduce the possibility of slipping if, for example, the car is driven in muddy gumboots.

The instruments are grouped in the middle compartment of the facia, and in addition to the speedometer there is an ammeter and a fuel level gauge. As well as the red light for ignition warning, lights are also provided to indicate low oil pressure (green), and head lamp main beam (red), while an amber light is illuminated if the mixture control is left out after the engine has reached its normal temperature. A combination switch, placed between the two circular instrument dials, controls the ignition and lighting, with a hand operated dip switch in the centre of the steering column.

The choke, starter, and the fan switch for the heater—if this is fitted—are placed below the facia panel, while the switch for the windscreen wiper is mounted on the back of the wiper motor, which is housed on the lower side of the windscreen wiper frame. Only one wiper blade is fitted as standard equipment, although an additional one can be supplied as an optional extra. The wiper itself is powerful, but it would be better if the blade cleared more of the upper part of the screen.

Semaphore direction indicators are fitted at roof level, and these are operated by a switch attached to a bracket on the steering column. The switch is convenient to operate, but it would be advantageous if it were either self-cancelling or fitted with a warning light. The test car was fitted with a heating unit mounted within the car, to the left of the central tunnel. This was effective both for heating the interior and demisting the windscreen, although the unit was rather noisy when the fan was in operation. In addition to the heater, four ventilators are provided, two in the scuttle and two in the roof panel.

The front doors have two-piece side windows, arranged so that the rear portion slides forward to increase ventilation or permit hand signalling. The two doors allow easy access to the front compartment, and the three front seats are of ample proportions, though the central tunnel restricts the leg room in the middle. Access to the rear compartment is by way of a door in the middle of the rear panel, and a folding step is attached to the rear cross member. The rear seats are attached to the sides of the body so that passengers in the rear face inwards. This arrangement means that it is possible to get to all four seats through the single door, and at the same time enables the seat cushions to be hinged, so that they fold up to leave a clear floor space when the rear compartment is used for the carriage of goods. There is very good all-round vision for the rear passengers, and, to improve visibility, rectangular Perspex lights are placed in the roof panels above the side windows and below the outer section of the double roof.

Although Rover saloon cars have a reputation for being particularly quiet, this is not shared by the Land-Rover which is, of course, a very different type of vehicle, and there is a noticeable amount of mechanical noise, particularly when driven in the indirect ratios; but bearing in mind the purposes for which the car is designed this is accepable. Although both head and tail lights are very good, the horn is barely adequate.

Under normal road operating conditions eight lubrication points require attention with an oil gun every 3,000 miles. The frequency of oil changes for items such as engine and transmission depends very much on the type of work for which the car is used. For example, if it is often engaged in deep wading, or operating in very dusty conditions, frequent oil changes may be desirable.

The Land-Rover Station Wagon is an outstanding car which can be driven almost anywhere. In addition to its suitability for cross-country work, it will put up commendable average speeds on normal roads. In Station Wagon form it provides satisfactory transport for up to seven persons while, if only three are carried, there is ample luggage space. The vehicle is ideally suited to towing a caravan or a horse box, it is robust and functional, completely free of unnecessary frills, and has the type of finish which is in keeping with the purpose for which it was designed. It is a first-rate machine of which the engineers of The Rover Company may be proud.

A short, flexible pipe connects the large air cleaner to the down-draught carburettor. The electric fuel pump is mounted on the bulkhead to the right of the electrical regulator unit, while the coil is placed on the left side of the bulkhead close to the distributor, which is carried high up towards the rear of the engine. The battery is in an accessible position towards the front of the right wing valance, by the side of the radiator

FULL INSTRUCTIONS for making

a simple, realistic toy

LAND ROVER

CUTTING LIST

		Long ft.	Long in.	Wide in.	Thick in.
(A)	2 Sides ..	1	2	2⅛	⅜
(B)	2 Wings (middle pieces) ..		4⅝	—	⅞ squ.
(C)	2 Wings (inner pieces) ..		4⅝	1¾	⅜
(D)	2 Wing fillers		1	1	⅜
(E)	1 Radiator ..		3⅜	1¾	⅜
(F)	1 Dashboard..		6⅝	2¾	⅜
(G)	2 Seat backs..		5⅞	1¾	⅜
(H)	2 Side seats ..		5¼	1⅛	⅜
(I)	2 Side seat supports ..		5¼	1	⅜
(J)	1 Main seat ..		5⅞	2	⅞
(K)	2 Rear (side pieces) ..		1½	2¼	⅜
(L)	1 Rear flap ..		3⅜	2⅝	⅜
(M)	1 Bonnet ..		5⅞	3⅞	⅜
(N)	2 Chassis members ..		10¾	—	⅞ squ.
(O)	2 Bumper extensions..		3¼	1	⅜
(P)	1 Centre chassis bearer ..		5⅞	1	⅜
(Q)	1 Rear chassis bearer ..		4⅝	1	⅜
(R)	1 Bumper ..		6⅝	⅝	⅜
(S)	3 Windscreen uprights ..		1⅛	⅝	⅜
(T)	2 Windscreen horizontals		6⅝	⅝	⅜
(U)	1 Cover top ..		9	6½	⅞
(V)	2 Cover sides..		9¼	2¾	⅜
(W)	1 Cover back..		5⅞	2¾	⅜
(X)	1 Cover front		5⅞	1⅜	⅜
	Hardboard ..	1	1	5¾	⅛

Allowance of ¼ in. has been made in lengths, ⅛ in. widths. Thicknesses are net.

The steering wheel requires a 1½ in. square of hardboard and a short length of ¼ in. dowel for the column.

Also required are a pair of 1 in. thin brass hinges, a pair of ⅞ in. dia. gliders, four drawing pins (two with red heads, two white), brass countersunk screws, four 3 in. dia. wheels, nuts, washers, and bolts. If rubber-tyred wheels as shown in Figs. 1 and 2 are difficult to obtain, wooden wheels can be shaped with a fret or bow saw from ¾ in. thick flat stuff. A rubber-tyred effect is produced by planting on a plywood ring and rounding the edges. Alternatively those having a lathe can turn them.

REPRINTED FROM
WOODWORKER
DECEMBER, 1955

*Trucks of all kinds appeal to youngster
special fascination. This toy version of
removed provides an ideal transport tr
easy to make. Only two thicknesses of
hardwood. The parts are put together
cross-pinning re*

FIG. I. A STRONG AND ATTRACTIVE TOY WHICH WILL PROVIDE ENDLESS ENJOYMENT FOR A YOUNGSTER

TO prevent splitting, guide holes for the nails should be drilled.

Body.—Prepare the two sides (A) and lay them side by side, ends level. By putting the point of the compasses in one piece you can strike the 3 in. diam. wheel arches on the other. The centres of these arcs are $\frac{1}{2}$ in. from the touching edges, $1\frac{3}{4}$ in. from one end for the front wheels and 3 in. from the other end for the rear wheels. Cut out the wheel arches and trim off the $\frac{1}{4}$ in. strips (Z), but do not cut off (Y) at this stage.

Prepare radiator (E) and cut off $\frac{3}{8}$ in. squares at the two lower corners. Nail an inner wing (C) to each side of the radiator and gently ease the pieces apart again leaving the nails projecting from pieces (C).

The front wings are assembled, making sure that the front edges are level. With a file, round over the solid front corner and the top and front outer edges, stopping where the dashboard will fit. Cut off (Y) level with wing filler (D). Finish with glasspaper.

Prepare dashboard (F), Fig. 3. The ends must be stepped to allow it to lap over the sides (A). A $\frac{1}{4}$ in. hole is bored to take the steering wheel column. This enters the dashboard $1\frac{3}{8}$ in. from the right hand end and 1 in. from the top. A $\frac{1}{8}$ in. pilot hole will form a guide for the $\frac{1}{4}$ in. drill.

Replace the radiator (E) between the two wings and add the dashboard (F), nailing through both sides (A) and also through (F) into the ends of the middle wings (B). The bonnet (M) should be rounded with the file and nailed in position as indicated.

Seats.—Prepare pieces (G, H, I, and J). Stand each side-seat support (I) on its edge and nail to each a seat (H), keeping the ends level and the edge of (H) flush with the face of (I). They are fixed to main-seat back (G).

Two shallow grooves are worked in the top and front edge of main-seat (J), giving the appearance of three seats. (J) is attached to (G) by two or three nails.

The seat back (G) is added after planing the lower edge at an angle so that it makes a snug fit with the seat when sloped backwards. The upper edge and front face is grooved to conform with the seat (J).

The whole unit is fixed between the sides (A). The upper edges of seat back (G) and the two sides (A) must be level, the whole being set in from the back $\frac{3}{8}$ in. to allow for three parts which comprise the rear.

Tailboard.—The three pieces (two sides (K) and flap (L) are shaped along their lower edges before fixing. Flap (L) is pivoted on nails between the rear sides (K). To ensure flush upper edges, measure the position of the pivots from the top of all pieces. Mark a point on the inner edges of pieces (K) and both edges of (L) $1\frac{5}{16}$ in. from the top. Drive panel pins half their length into the marks on (L) and drill corresponding holes in pieces (K). Nip off the nail heads and fit the three parts together. The curve of the lower edges can be worked and the complete unit secured between the sides of the body. To complete the body, add the hardboard bottom (cut to the shape indicated in Fig. 4) using $\frac{1}{2}$ in. panel pins.

Chassis.—Main members (N) are drilled to receive the axles. The axle centres are $\frac{1}{2}$ in. from the top edges, $\frac{7}{8}$ in. from the rear ends and $\frac{5}{8}$ in. from the front ends. Mark both sides of each member, and drill the axle holes half-way through from each side. The size of these holes depends on the type of wheels used.

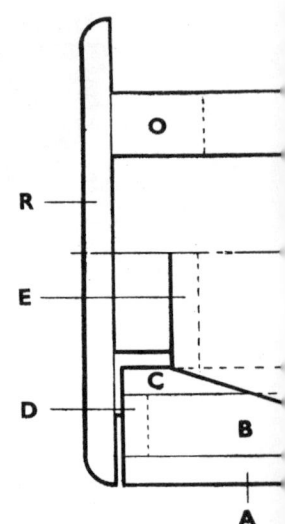

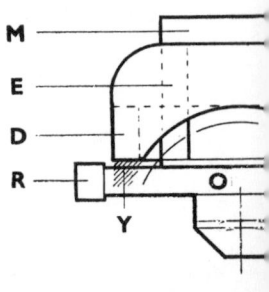

FIG. 3. PLAN, ELEVATIO

OVER • •

...hat looks just like the real thing has a
...ver has a detachable hood which when
...en bricks, pebbles, and the like. It is
...equired ; preferably a straight-grained
...nd glue. The method of assembly and
...g construction

FIG. 2. WITH THE HOOD REMOVED IT CAN BE USED FOR CARRYING CHILDREN'S BRICKS AND THE LIKE

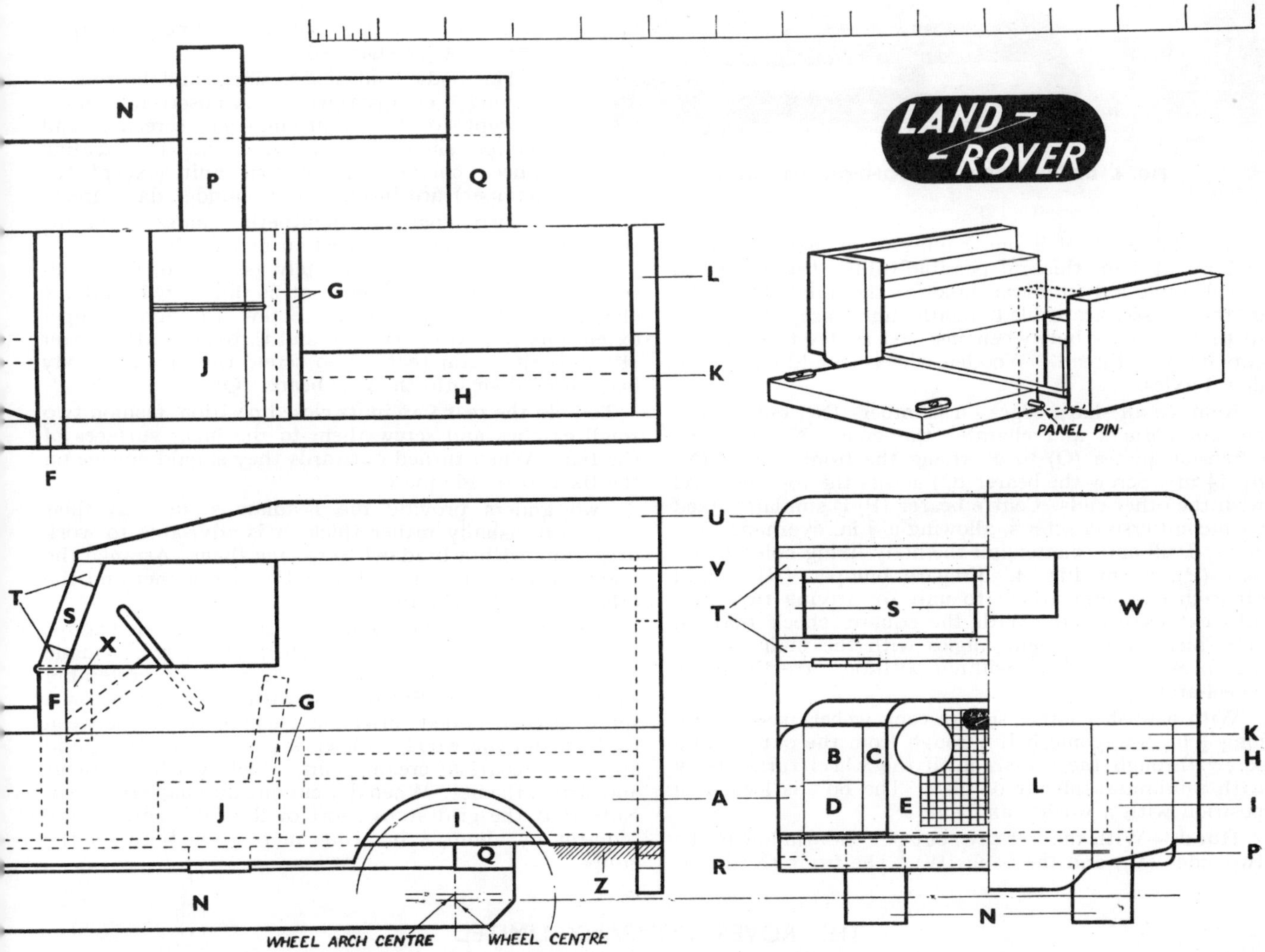

LAND-ROVER

PANEL PIN

N   P   Q

G   L   J   K   H   F

T   S   X   F   G   J   I   Q   N

U   V   T   S   W   B   C   D   E   L   A   R   Z   N   K   H   I   P

WHEEL ARCH CENTRE   WHEEL CENTRE

CTIONS. Top right, the Land-Rover sign roughly to scale which can be cut out or copied and fixed to the radiator grille. Below, method of hingeing the tail-board. Radius of wheel arch, 1⅝ in.

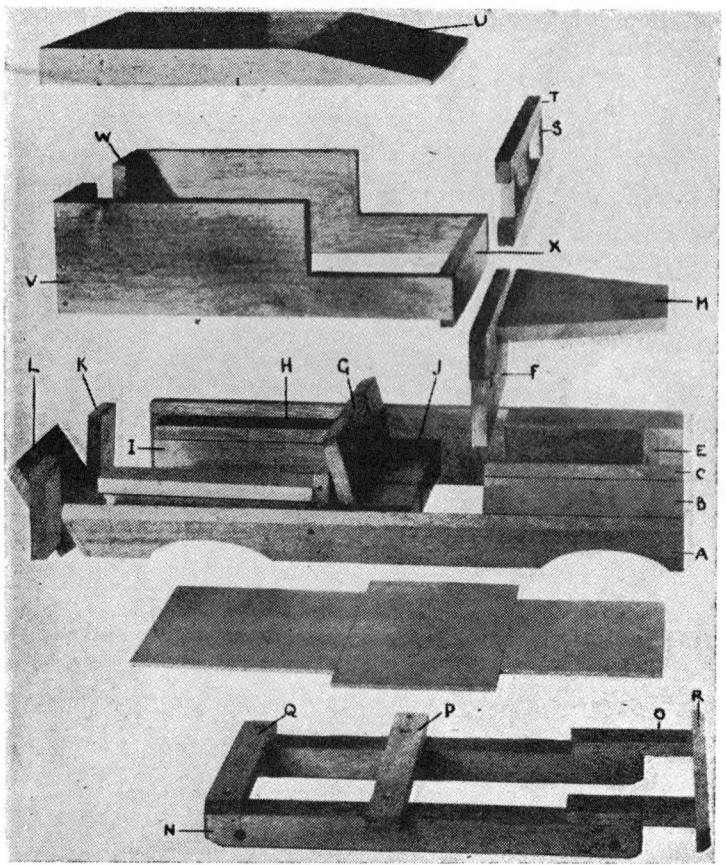

FIG. 4. DETAILS OF MAIN COMPONENT PARTS

to fit over the steering wheel column. These four parts are glued and nailed together. Note that the front is stepped down to provide positioning flange for the hood when fitted on the body (the back of the main seat acts as another).

The front slope of the top (U) is obtained by planing. Before rounding over the top edge of each side, drill the holes for the nails and attach the top to the rest of the unit. Punch the nails at least $\frac{1}{4}$ in. below the surface so that they do not foul the file when rounding over.

Windscreen and Steering.—The windscreen consists of three uprights (S) positioned between two horizontals (T). Allow a generous overhang on the horizontals to prevent any tendency for the wood to split when nailing. Punch nails well in so that the plane does not foul them when working the bevels. Plane top and bottom to the angles shown until the windscreen makes a snug fit between dashboard and hood. Trim off the overhang and glasspaper smooth.

The steering wheel is a circle of hardboard screwed to a length of $\frac{1}{4}$ in. dowelling. Do not fit windscreen or steering until all painting is completed.

Finishing.—Fill nail holes and blemishes, and rub down all surfaces with fine glasspaper, removing the sharp edges.

The body load space, sides, and bottom of driving compartment, windscreen, hood (except front and driving side screens), and bumper should be painted cream. The steering wheel, column, and chassis unit (except the bumper) are black, the remainder dark green.

Two coats of synthetic enamel will be sufficient as this does not need an undercoat.

The toy can now be completely assembled, the windscreen being hinged at the front as illustrated, and the chassis fixed to the body. Screw through bumper extensions (O) into parts (C), and through centre bearer (P) into the main seat. Also screw through the body load-space floor into the rear bearer (Q).

To hold the rear flap in its closed position, fashion two small catches and screw them to the inner surfaces of the flap. When turned outwards they should engage on the fixed rear side pieces.

Two gliders provide the headlamps and, as their points are usually rather thick, it is advisable to work little slots with a bradawl to receive them. Arrange the points so that one enters the end of the bonnet and the other two enter the radiator.

Two white-headed drawing pins provide the side lamps and red-headed ones the rear. The radiator grid is made from a 4 in. square of plain net curtaining—preferably with a $\frac{1}{8}$ in. mesh. Stretch out with drawing pins on a spare piece of wood, inserting packing pieces between the net and the wood, to keep it clear of the base, and apply two coats of cream paint, which will both colour and stiffen the net. When dry cut out the shape required. Note that the grid stops short of the full depth of the radiator by about $\frac{1}{4}$ in. Glue in position between the lamps.

A suitable wheel is one having a diameter of 3 in. and about 1 in. thick, if possible rubber tyred. Metal wheels have a $\frac{1}{4}$ in. diam. hole for the axle. The holes in the chassis are drilled slightly undersize, say $\frac{1}{32}$ in., so that the axle bolt when inserted as described below cuts its own thread. Wooden wheels should have $\frac{3}{8}$ in. dowel axles.

Remove all sharp edges on members (N), cut off the bottom corners and chamfer the ends. Nail on the extension pieces (O) to overhang the front end of (N) by $1\frac{1}{4}$ in. Screw the bearer (Q) across the top and level with the other ends. Centre bearer (P) is similarly fixed with countersunk screws, allowing a $\frac{5}{8}$ in. overhang each side. Drill and countersink the body fixing holes in (O) and (P), as in Fig. 4. Prepare bumper (R), round off each end, and attach to unit by driving two nails into extensions (O). With the square, check that all cross-pieces are at right angles with the main chassis members, or the wheels will not position correctly in the wheel-arches.

With a washer either side of each wheel, pass a 2 in. long $\frac{1}{4}$ in. hexagonal bolt through from the outside and screw through the chassis until the wheel turns freely with minimum amount of play. The bolt is locked in position with a washer and nut.

Hood.—With the fret saw, cut out the window in the two sides (V), and the back (W). The front (X) is cut

THE ROVER COMPANY LIMITED
SOLIHULL - BIRMINGHAM - ENGLAND

ROUGH GOING—this is on Sir Alan MacLean's Littlewood Park estate, Alford, Aberdeenshire—causes the Land-Rover no trouble.

READY FOR ANYTHING
By S. C. H. DAVIS

The Land-Rover, practical and without frills, can tackle any ground and be used in ways that were never intended

THOSE who have had an engineering training know exactly what results from being regarded as an "odd job man" from whom mechanical miracles are expected should anything in the village go wrong. In the vehicle world the Land-Rover must feel much the same, for this very useful little car is able to tackle almost anything and, with a little ingenuity, can be made to do things its designer never contemplates, and do them with a will.

In normal form it is a sturdy machine, essentially practical, without frills, made to stand that neglect and mishandling which is the inevitable lot of its kind, and most entertaining when tackling apparently impossible terrain. On the land between Bisley and Aldershot there is a stretch often used for weird manœuvres with W.D. vehicles and exuberance by motor-cyclists happily developing their skill in riding across country but, at sight, horrific to the owner of the ordinary car. On it the Rover took everything in

---

SPECIFICATION
Engine: Four cylinders, 77·8×105 mm., 1,997 c.c. R.A.C. rating 15 h.p.
Gear ratios: High, 16·171, 11·026, 7·435, 5·936 to 1; low, 40·688, 27·742, 18·707, 13·578 to 1.
Weight 1 ton 4½ cwts. Length 11ft. 9in., width 5ft. 2 $\frac{9}{16}$ in., height 6ft. 4in. Ground clearance 8in. Turning circle 37ft. Fuel tank, 10 gallons.
Price £570 (no tax).
Extras: Power take-off behind gear box, £8. Winch £40. Governor £15 15s. Extra mirror 15s. Heater £8 15s. Turn indicators £2 15s. Rear power take-off £27.

---

its stride, including soft-surfaced gradients so steep that the machine had the air of a fly crawling up a wall, while the relative angles of the wheels over ruts, ditches, and mounds were fantastic.

All this, of course, is due to having eight gears and four-wheel drive whenever you want it, steering suitable to the work, and an engine which can pull happily at low speed. In contrast it can take seven people anywhere, three of them in front, and cruise at an indicated 50 m.p.h., with a maximum of well over 60 at need.

The amount that can be stowed in the back is surprising, varying from animals to packing cases, in which it brings back memories of those French rail trucks so thoughtfully labelled "eight horses, forty men." It is not quiet, nor softly sprung, yet much more comfortable than Fi-fi the Jeep which was my constant and essential companion in the late scuffle over Europe. Incidentally, you sit above the fuel tank in just the same way, but there being no enterprisingly hostile folk about with Mausers you do not get the same tingling down the spine when passing through woods.

The Land-Rover has a sense of humour also, liking to rattle chains unexpectedly in a way similar to the inhuman inhabitant of the moated grange, and once it created considerable excitement by setting up a hair-raising shriek which was traced, after great trouble, to the wind playing with the minute opening left because a door had not been closed properly.

With the hood up you want more than the one outside mirror to see properly astern, and the blind spots on the quarters need watching, but the car remains dry in a torrential downpour and is reasonably warm. To me the brake lever was awkwardly placed, though it was entirely effective. Brakes generally are good and for what it is worth, the machine will get up to 50 m.p.h. from 30 in 12 seconds, and accelerates with a rush on third using the higher set of ratios.

Everything, naturally, is austere, instruments included; but there are good shelves, practical ventilators, and the wiper clears the screen over a wide arc, while the engine auxiliaries are accessible.

Provision is made for a power take-off, at the rear or amidships, and a winch with which you can do practically anything. The towing attachment is strong and easy to adjust, and the spare wheel can if necessary be carried on the bonnet, thus clearing the back of the body still more. There are two types of wheel and tyre available, according to the work the car has to do, and one type of wheel has a split rim, making it easier to remove or fit the larger tyres. (Doing so, remember what happened when the inquisitive but ignorant soldiery undid the relevant nuts with the tyre fully inflated, thereby producing an explosion of atomic proportions and no small damage to themselves.)

For the machine there is a wide choice of bodies and an even wider selection of accessories, though purchase tax creeps in if you select the most luxurious equipment. A heater, extra wiper, mirrors, demister, hand throttle, governor, softer upholstery, and turn indicators are available, and as for bodies the machine can be a truck, van, farm vehicle, fire engine, or travelling welding plant.

For jobs of every kind on farms and on big estates, as a tender, or a tug, this is one of the most useful vehicles ever produced, and it would go anywhere on those primeval continents to which by tradition go disappointed suitors with a view to forgetting fair ladies by pursuing unintelligent pachyderms.

By G. S. SHARPE,
A.M.I.Mech.E., M.S.A.E.

Larger body increases potentialities of the versatile 15 cwt Land-Rover

GO-ANYWHERE PICK-UP

(*Left*) *Through a watersplash during the test of the pick-up Land-Rover.*

NO other vehicle is as widely employed for such a variety of purposes as the Land-Rover. Its four-wheel-drive with high- and low-speed ranges makes it capable of traversing almost any kind of terrain and the power take-off drives can be utilized to transform the vehicle into anything from a mobile welding plant to a fire-tender or crop sprayer. But some potential users of the Land-Rover principally for transport duties have in the past been critical of the small load-carrying space for its rated 10cwt. capacity despite the fact that it has sufficient power to tow 20 tons or more on the flat without difficulty.

It was for this reason that the wheelbase was increased about a year ago. The change resulted in a 25 per cent increase in body space and a new and enlarged pick-up version with a capacity of 15 cwt. was put into production. A road test of this larger version has shown that the additional weight has had a very slight effect only upon the performance and that the Land-Rover has lost none of its characteristic ability to go anywhere on or off the road Incidentally, the popularity at home and abroad of the Land-Rover of both types can be gauged from the fact that the present production rate of 500 per week is insufficient to meet demand and factory extensions now under construction will provide for an output of 1,000 per week.

A number of modifications and improvements made to the Land-Rover at the time of its increase in size were also incorporated in the 15 cwt. pick-up truck and the mechanical specification of the two models is broadly the same. The engine is the well-stabilised four-cylinder, 2-litre Rover petrol unit with overhead inlet and side exhaust valves. It develops 52 b.h.p. at 4,000 r.p.m. and has a maximum torque of 101 lb.-ft. at 1,500 r.p.m. The clutch is a single dry-plate unit of 9 in. diameter and a four-speed gear

box is unit-mounted with the engine. Immediately behind it is a two-speed transfer box incorporating a two- or four-wheel-drive control on the output shaft. The gear box has single helical constant mesh gears with synchro-mesh on top and third and the ratios are first, 2.996; second 2.043; third, 1.377 to 1; and top, direct. For normal road work the transfer box high ratio of 1.148 to 1 is used and with the spiral bevel front and rear axle ratio of 4.7 to 1 gives overall reductions of 16.17, 11.02, 7.43 and 5.93 to 1 in top.

Rear-wheel drive only is generally employed on the roads, although a control is provided to engage the front-axle drive when required. When the control lever for the transfer box is placed in the low ratio position it brings in the reduction ratio of 2.88 to 1 and also automatically engages four-wheel drive. The overall final drive ratios in the low transfer ratio are indeed so low that it can safely be assumed that where they are required, then four-wheel-drive is also necessary. In low top gear the overall ratio is in fact very close to that of first gear in high transfer. The actual low ratios are first, 40.46; second, 27.72; third 18.70; and top, 13.57 to 1.

Changes from the low to the high ratio of the transfer box can be made while the vehicle is moving at any speed and the change automatically disengages front-wheel-drive. This engagement thus ensures that on normal roads the drive is usually through the rear wheels only. From the

tyre wear aspect this is, of course, to be desired on dry roads. On snow and icebound surfaces, however, a drive on all four wheels can be as great an asset as on wet muddy, off-the-road-tracks—hence the separate control provided to engage front-wheel-drive when required in the high ratios.

Unladen the pick-up weighs 1 ton 7¼ cwt. so that with a full 15 cwt. load as tested the gross running weight was 2 tons 2¼ cwt. A standard type Land-Rover of the short wheelbase type in production before these latest modifications weighed 1 ton 16 cwt. when tested in 1953 with its 10 cwt. rated load. The larger body, has not, therefore, caused much increase in unladen weight. The pick-up has a wheelbase of 8 ft. 11 in. compared with 7 ft. 2 in. on the latest type Land-Rover, and its light-alloy body has interior dimensions of 6 ft. long by 4 ft. 9 in. wide. The body is 1 ft. 8 in. deep at the centre for a width of 3 ft. and shallow benches 8⅛ in. high run the full length of the body on each side. These box in the wheel arches, provide useful lockers and form seats for carrying personnel when necessary.

The cab is fully enclosed, having a metal roof and a rear panel in which a wide two-panel sliding window is fitted. So far as driving comfort is concerned, therefore, the pick-up is fully weatherproof and its full width seat provides comfortable seating for two passengers in addition to the driver. It was, in fact, with the additional weight of two passengers that the pick-up climbed a wet grassy bank with a gradient of 1 in 2½ during the test. Even this did not call for the lowest reduction ratio available, but was done in low second gear. The four-wheel-drive is, of course, automatically engaged in the low set of ratios, but it is obvious that the vehicle will climb any gradient on which wheel grip can be maintained. There was only one occasion on which the pick-up failed to negotiate the same hazards that had been tackled in the earlier test of a standard Land-Rover. This was when attempting to reach higher ground from a farm track by way of an almost vertical bank about 2 ft. 6 in. high. The longer rear overhang caused the back of the body to foul when the front wheels reached the top of the bank. However, these were abnormally severe test conditions unlikely to be attempted in normal service.

The suspension stood up well to a great deal of rough treatment during the off-the-road trials and is obviously designed to meet these conditions. Consequently, it provides a very firm ride on hard roads, but is still sufficiently flexible to enable long journeys to be undertaken without undue discomfort or fatigue. Resulting from this there is also an absence of roll and a high-geared steering combines to give a degree of control in fast cornering which is appreciated. The handling varies in the unladen and laden conditions between slight understeer and slight oversteer

Girling hydraulic brakes are fitted and the pick-up has drums of 11 in. diameter, which is 1 in. greater than the standard version of the Land-Rover. The shoes are leading and trailing on both front and rear, and tests on the road showed that powerful braking was available at all times. Figures as high as 90 per cent. efficiency were recorded on the Tapley meter when stopping from 30 m.p.h. and very little fade was experienced. Furthermore, after wading, the recovery was as quick as could be expected. The hand brake is on the transmission and is of the drum type mounted on the gear box output shaft. It is extremely effective and proved capable of holding the vehicle on the 1 in 2½ gradient climbed during the test. It is intended as a parking brake only and although it can be

used to stop the vehicle, this is not an advisable or effective method. The foot brake, as has already been shown, provides the maximum braking possible.

There is a de-luxe version of the pick-up which has a number of refinements in the cab fittings including linings to the roof, rear panel and doors. There is also a one-piece back-rest for the three separate seat cushions and plastic-covered felt mats for the floors. A number of optional extras are also available for both the standard and de-luxe models and include heater and de-mister, tonneau cover or a hood for the body, arm, front capstan winch and a tropical cab for roof.

Quite apart from the unique performance characteristics of the Land-Rover it has the particularly attractive feature of being exempt from purchase tax. The pick-up in standard form costs £635 and in view of its usefulness either as a general-purpose vehicle, or as one used for a particular application, it represents exceptionally good value for money.

On a steep grassy gradient more severe than it appears— actually 1 in 2½.

THE ROVER CO. LTD
SOLIHULL
BIRMINGHAM

respectively, so that it would appear that the best compromise has been attained. The front-wheel-drive imposes a limit on steering lock and the turning circle diameter is 50 ft., but this is a small price to pay for the advantages obtained.

All controls are well placed and simple to operate. The gear change lever, front-drive control and transfer lever for high and low ratios are all centrally positioned as is a pull-up type handbrake. The clutch and brake pedals are of a good size and fitted with side flanges to form a more positive anchor for muddy boots. The instruments are grouped centrally on the facia panel and a useful tray provided on each side. Engine-starting from cold is particularly good even after a night in the open with temperatures below freezing. The warm up also is rapid and little time is lost in getting under way. The low gearing ensures a particularly lively acceleration from rest through the gears and in top gear also the performance is very flexible. The close ratio-gear box is very pleasant to handle and very good acceleration times were obtained. From rest to 30 and 40 m.p.h. the average times from runs in two directions were 10.8 and 19.0 sec. respectively. In top gear 30 m.p.h. and 40 m.p.h. were reached from 10 m.p.h. in 13.7 and 23.7 sec. respectively. These figures were taken with the rear wheel drive only engaged and were, therefore, in the high ratio of the transfer box. In the unladen condition, the Land-Rover is, of course, even more lively and the times to reach 30, 40 and 50 m.p.h. from rest were 7.65, 13.3 and 22.2 seconds. The low gearing limits the maximum speed of the vehicle to 60 m.p.h., but this is more than adequate for a vehicle of this type.

Long-distance runs can be made at fast average speeds if necessary and the petrol consumption can be kept down to a very economical figure without difficulty. A careful check was made under good main road conditions and fully laden the vehicle gave a figure of 21 m.p.g. at an average speed of 37 m.p.h. Off the road the consumption obviously increased depending upon the actual running conditions, but there is every indication that the overall efficiency of the Land-Rover is high

Wide doors hinge right forward or may easily be removed altogether.

Despite the spare wheel position on the bonnet top, engine accessibility is good.

Easy loading with the aid of the tailboard and provisions for towing are features of the vehicle.

THE LAND·ROVER

This month's "Car from Britain" is probably nearer a commercial vehicle than any other model so far considered in these pages. All the same, its owner is as likely to be a private individual as an organisation.

The Rover company, who make this excellent machine, enjoy a high reputation as automobile builders; and there is nothing about this work-a-day little wagon that detracts from their good name. There is no denying that a well known military vehicle, built across the Atlantic, was the ancestor of the Land-Rover; but as it stands to-day it is as far from the "Jeep" as we are from Neanderthal Man.

It is difficult to know where to begin a description of the car, which is available with so many "extras" as to be in effect several different cars. Basicly it is a four-wheel drive chassis, available in two lengths, with a four-cylinder engine of about two litres capacity. All models have two complete sets of gear ratios – one for use on the road and one for "cross country" – and all can seat three across the front seat.

There are endless variants of the rear half which will accommodate anything between, say, three pigs and eight children. Although "open" in its basic form, the Land-Rover can be fitted with a very adequate canvas top or, on demand, a metal cab (removable in parts, and interchangeable with the canvas). The back ends can be covered in canvas or metal – again with interchangeability. The metal cab can be supplied with a Tropical Top. A winch can be fitted between the front dumb-irons, and there is provision for a power take-off at the rear.

On the other side of the picture, the metal cab, as provided on the model loaned to THE AMBASSADOR, is very comfortable. The seats, covered with "Tygan", are smart and soft. It can be fitted with a heater and, if required, a radio. Here the true Jekyll and Hyde character of the vehicle is revealed; for, having taken the pigs to market, sawn a little wood, and driven a hundred miles through a quagmire, the car can be used to take two girls in full evening regalia to a dance – with comfort whether in the Tropics or in the Arctic Circle.

Having said all this there remains only the question of performance, which must be taken in two parts – on the road and very much off the road. On the road, using the rear wheel drive only, there is nothing about the Land-Rover which would lead a novice to believe he was driving anything but a normal car. The long wheelbase model (pictured here) cruises at about 45/50 m.p.h. along main roads. Its brakes and steering are well up to standard, and a petrol consumption of a little more than twenty miles to the gallon cannot be called excessive. The actual car we handled was inclined to "pink" on any but the best grade of fuel; but, in view of the astonishing performance of which the car is capable, a demand for the best fuel is not unreasonable. Off the main road the car's performance becomes fantastic. We drove it for some miles through open moorland, and across considerable ditches. We finally brought it to a halt by driving up a hill which had been used for testing tanks – the gradient was about one in two and the surface deep loose sand. Even here we got within a few yards of the top of a two hundred yard climb. It is only a slight·exaggeration to say that a Land-Rover will go anywhere a tank will go.

Of course the car is not for everyone; but for farmers, backwoodsmen, and any others who need a car which can be used both for work and for pleasure, the Land-Rover in one guise or another provides facilities unobtainable in any other vehicle.

The photograph below brings out one side of the Land-Rover's character. The gun is a 20-bore, with pistol grip and 26-in. barrels, weighing 5 lb. 7½ oz. Its oak and leather case, which also holds cleaning materials and accessories, is seen (opposite) on the front of the Land-Rover, together with a 100-size pigskin cartridge bag. The makers are JAMES PURDEY & SONS LTD., LONDON W.I.

James Purdey & Sons Ltd. also supply the lightweight waterproof in an imperishable rubberised Irish linen, manufactured by LEE BROS. (OVERWEAR) LTD., LONDON W.I, *with two side pockets with buttoned flaps, two large game pockets inside, and buggy lining over shoulders. Hooks and eyes are fitted so that the skirt of the garment can be hooked up to make a deep wading jacket.*

LAND-ROVER: SPECIFICATION

ENGINE Number of cylinders, four; bore, 3.063 ins. (77.8 mm.); stroke, 4.134 ins. (105 mm.); cylinder capacity, 121.9 cu. ins. (1,997 c.c.); maximum b.h.p., 52 at 4,000 r.p.m.; maximum torque, 101 lb./ft. (14 m./kg.) at 1,500 r.p.m.; compression ratio, 6.8 : 1; three-bearing crankshaft, overhead inlet and side exhaust valves, hydraulic camshaft chain adjuster, pressurised radiator, oil pump intake and by-pass oil filters, and oil bath air cleaner.

CLUTCH Single dry plate, 9 ins. (230 mm.) diameter.

DYNAMO Automatic voltage regulator, 12 v. **STARTER** Operates on flywheel.

CARBURETTOR Downdraught. **TYRES** 7.00 × 16.

BRAKES Hydraulically-operated foot brakes. Mechanically-actuated handbrake operates on transmission shaft to rear axle.

REAR AXLE Semi-floating, spiral bevel type. Ratio, 4.7 : 1.

FRONT AXLE Fitted with differential similar to rear axle. Drive to front wheels through totally enclosed universal joints.

COOLING SYSTEM Thermostatic pump and fan; capacity, 17 pints (9.75 litres).

LUBRICATION By pressure from gear type pump forcing oil to all bearings, timing chain and valve gear. Capacity, 10 pints (5.5 litres).

TRANSMISSION Provision for eight forward speeds and two reverse, and two- or four-wheel drive by means of transfer gearbox.

Main Gearbox		Transfer Box	
		High Ratio	Low Ratio
First gear	..	16.171	40.688
Second gear	..	11.026	27.742
Third gear	..	7.435	18.707
Top gear	..	5.396	13.578
Reverse gear	..	13.745	34.585
	Final drive ratio, 4.7 : 1		

Four-wheel drive is fitted for optional selection when high transfer gear is in use but is engaged automatically when low transfer gear is in operation.

IGNITION Coil and battery, automatic advance. Battery, 12 v., 51 amp./hr.

CHASSIS Side and cross members of box section forming exceptionally rigid assembly.

STEERING Worm and nut, 15 : 1 ratio. Right- or left-hand steering as required.

FUEL SUPPLY 10-gallon (45 litres). **PAY LOAD** 1,500 lb. (680.4 kgs.).

SPRINGS Semi-elliptic front and rear. Telescopic type shock absorbers front and rear.

DIMENSIONS Overall length, 173½ ins. (4.41 m.); overall width, 62⅞ ins. (1.59 m.); overall height, 78 ins. (1.98 m.) (unladen) without rear hood; overall height, 83½ ins. (2.12 m.) (unladen) with rear hood; wheelbase, 107 ins. (2.72 m.); track, 50 ins. (1.27 m.); ground clearance, 8¾ ins. (222 mm.); turning circle, 48½ ft. (15.25 m.); unladen weight 3,056 lb. (1,400 kgs.).

REAR POWER TAKE OFF (At extra cost.) Drive through back of main gearbox to rear of chassis. Can be fitted to give pulley drive for threshers, chaff cutters, circular saws, etc., or shaft drive for power mowers, binders, combine harvesters, etc.

BODY Body panels of non-corrodible light alloy, external steel fittings galvanised.

OPTIONAL EXTRAS The following is a selection of some of the optional extras available on both standard and de Luxe models: hood covering rear section, with windows front and rear; rear power take-off; heater and demister; tropical roof for cab; rear seats and backrests; front capstan winch; trafficators; tonneau cover for rear section; radio; extra windscreen wiper; canvas spare wheel cover.

PHOTOS BY HEPHER

"Operation Roverland"

A 12-month Trip by Land-Rover from Hobart to London Through 12 Countries

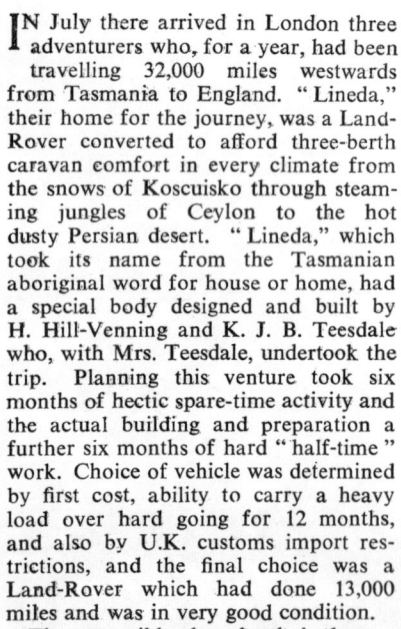

Everything *including* the stove. This was the Land-Rover's load at the start of the trip from Tasmania. Note how the roof lifts to provide room for two upper berths protected by canvas side-screens. The trailer was dispensed with before leaving Australia.

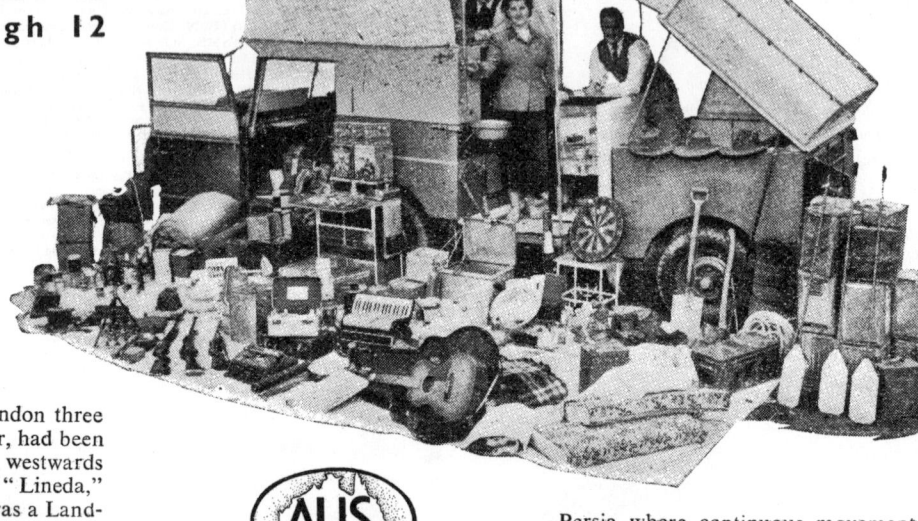

IN July there arrived in London three adventurers who, for a year, had been travelling 32,000 miles westwards from Tasmania to England. "Lineda," their home for the journey, was a Land-Rover converted to afford three-berth caravan comfort in every climate from the snows of Koscuisko through steaming jungles of Ceylon to the hot dusty Persian desert. "Lineda," which took its name from the Tasmanian aboriginal word for house or home, had a special body designed and built by H. Hill-Venning and K. J. B. Teesdale who, with Mrs. Teesdale, undertook the trip. Planning this venture took six months of hectic spare-time activity and the actual building and preparation a further six months of hard " half-time " work. Choice of vehicle was determined by first cost, ability to carry a heavy load over hard going for 12 months, and also by U.K. customs import restrictions, and the final choice was a Land-Rover which had done 13,000 miles and was in very good condition.

The extensible three-berth body was designed after consideration and rejection of tent living, hotels and caravanning. The first on grounds of inconvenience in bad weather and hard terrain; the second on expense, non-availability and necessity to make and keep bookings; the third due to difficulty of towing over creek crossings, etc., and expense when shipping, since four sea crossings were necessary: Tasmania-Australia, Australia-Ceylon, Ceylon-India and Continent-U.K. When extra fuel and water tankage and refrigerator were decided on, a trailer became necessary and it was designed to fit on

the bonnet for shipping, to occupy space paid for but unused in normal cases. The extra tankage was later found unnecessary and the trailer was, in fact, disposed of in Australia.

Novel Features

The body of the Land-Rover was built of welded seamless conduit panelled with aluminium sheet, felt-insulated and lined with leather-covered masonite. When the roof is raised over the two upper Dunlopillo berths, canvas and mosquito net walls unfold for weather protection. Similar canvases, carried loosely, fasten around the roof, body and tailboard edges to form a tent around the cooking and washing facilities. The rear compartment is fitted with lined cupboards and shelves for two-burner stove, pressure cooker, typewriter, medical kit, food and utensils, cine and miniature cameras, etc., and a shower and wash basin are mounted in the rear over the special large steel tailboard, which forms a firm floor when lowered on to levelling jacks. A hinged headboard behind the centre seat folds to form the end of a lower berth, usable whilst motoring. Some of the other interesting features are listed overleaf.

Roads on the whole proved fairly negotiable in the dry seasons, except in

Persia where continuous movement of heavy transport carrying large loads at high speed formed corrugations which punished the car severely. Rate of tyre replacement was fairly high, particularly in India where carts force vehicles off the narrow strip of concrete on to razor-sharp bullock shoes embedded in the soft sand. The highest price paid for a single 7.00/16 tyre and tube was the equivalent of £22 sterling, in Pakistan.

Both sand and mud proved to be excellent bogging-down media: once in Australia's Northern Territory trees had to be felled to use as levers and the assistance of aboriginal stockmen obtained to extricate the Rover from knee-deep mud. In North Queensland they were the rescuers when a Ford Utility had been caught by the tide and on another occasion they towed a 6-ton truck loaded with scrap metal! Persian sand bogged the Land-Rover twice, one digging operation being made particularly difficult by a flat tyre caused by a spinning wheel fouling an embedded rock.

Water, though brackish in outback Australia and Persia, was always obtainable, though sometimes from doubtful sources, and had to be purified by boiling or the addition of tablets. Canvas water-bags hung in the airflow through the folded windscreen provided cold drinking water in temperatures reaching up to 116° F. " Cool-a-Ride " type cushions are indis-

(*Above*) K. J. B. Teesdale and a young lady hitchhiker from London have a rest after helping to get the Land-Rover out of knee-deep mud in Northern Territory, Australia. (*Left*) The 27th puncture; this one was near Dalbandin, Pakistan.

"Operation Roverland" - - - - - - Contd.

pensable on a trip of this nature. Food was regarded philosophically; tinned reserves were kept to a minimum, cost and weight being great, and the majority of food purchased daily in the local markets. Some knowledge of French and German and a smattering of the local tongue overcame the language difficulties.

Driving through Eastern countries is not a very restful occupation and needs the aid of a powerful horn, preferably wired to the ignition switch! In India, bullock carts, herds of goats waiting their turn at the traffic lights, and pedestrians whose antics defy intelligent anticipation, share the roads with cars, cycles and rickshaws. Night driving is even more nerve-racking; instead of dipping, lights are totally extinguished for lengthy periods, leaving approaching cars to negotiate in complete darkness. Almost as disconcerting was the Persian method of flashing headlights at oncoming vehicles. Driving here is wild though expert and contrasts strangely with the almost over-careful manœuvres of the English motorist.

Turks seemed to specialize in turning cars and trucks over the mountain sides, for no less than seven were passed during a four-day journey.

Vehicle maintenance was of a high standard, particularly in Persia, and petrol, though frequently adulterated throughout the East, was readily obtainable. Cost varied from under 2s. per gallon in Persia to over 10s. in Pakistan, averaging 6s. per gallon throughout the trip.

The people of most countries were kind and hospitable; Indians in a gracious manner, Persians with a militarized gruffness, and the sturdy Turks with cheerfulness. Australia, particularly the North and West, offers a "boom" atmosphere; the Indian nation is awaking from a slumber of centuries, while the tired aspect of France differs widely from the evident vitality of Germany. In England one is struck by the neat dolls-house appearance of roads and houses as compared with the spaciousness of Australia.

This venture, though more costly and rigorous than sea travel, offered

exhilarating experiences and a much greater personal contact with peoples and cultures. To those who contemplate such a trip these three travellers offer encouragement and the benefit of their experiences.

They even know the whereabouts of an excellent vehicle fully equipped for the trip!

"THE ROVERLANDERS."

Notes and Equipment

Extra equipment and modifications to the Land-Rover included: Fitting Windtone horn, spare coil and condenser in ready-to-use position; lengthened and reshaped gearchange lever; tools and spares mounted under bonnet and in boxes on trackrod protection bar between dumb irons. Special extra six-gallon fuel tank mounted under centre seat fitted with change-over cock, and fuel and water tanks on top of "kangaroo guard" of welded square-section tube mounted above bumper. Hand-operated winch (¾-ton pull), axe, shovel, crowbar and tow chain were also mounted on this tubular member. Total tankage (car only): Oil, six gal., water 10 gal., fuel 32 gal. (including two jerricans). Windscreen cut and repivoted, doors re-hinged to enable screen to fold over bonnet-mounted spare wheel, thus allowing "pressurization" of vehicle interior to exclude dust. Wheel arches rubber lined and fitted with mud flaps; tubular rails fitted on roof under doors and over wings as tie-ons and hand holds. Experience dictated repositioning of fuel pump in cab to prevent vapour lock (pusher role).

Clayton heater/fan; 12V./240V. vibrator for electric shaver; 12V./240V. 7-valve radio.

Although not used throughout the journey as originally intended, the trailer showed ingenuity in design. For shipping, cost was saved by mounting it on the bonnet; for reconnaissance and ferrying, its lid became an entirely practical boat, complete with rowlocks.

WE TRY THEM OUT

By
C. S. SMITH
and
TOM GILLING

Probably one of the most useful vehicles for the farm ever built in Britain was tried out in our fifth road test and proved itself able to go anywhere and do almost anything

5. Land Rover

All this and a snowstorm too. The gradient is about one in three and the surface rutted, with the gravel and sand frozen.

THE official pointed on the map to a farm at the head of a valley in the Welsh mountains. "You ought really to go and see Mr. Pugh," he said, "but I don't suppose you'd be able to get there."

We said we had a Land Rover. "In that case you might," he replied, "but I warn you, it's very bad up there."

It was and we did. By that time there was snow and the hardest frost for years—even some of the mountain springs were frozen.

That, in effect, sums up the Land Rover. It will go anywhere under any conditions wherever wheels can grip—the top of a mountain, the middle of a desert, or through the Khyber Pass. Four-wheel drive, big wheels, eight gears, and the most rigid chassis frame ever built for this size of vehicle, will take you up or down fantastic gradients on surfaces where a mountain pony would have to watch its step.

This ability to negotiate anything, and yet to put up a highly acceptable road performance, gives a driver a very pleasant feeling of confidence. As well it can be used for harrowing, hay sweeping, driving (through p.t.o.) cutterbars, hedge-cutters, sprayers, dusters, compressors, hammer mills, winches or saw benches, which explains why the Land Rover when it was introduced in 1948 became immediately popular with farmers. Small wonder that some farmers we visited said, "You can leave that thing here."

The model tested was the standard 86 in. wheelbase, with the detachable canvas tilt—the general factotum of the range. It was fitted with dual purpose 600x16 in. tyres.

The test covered 1,200 miles in the worst weather of February, during the hardest frost for ten years, and under the most abominable road and off-the-road conditions.

Nearly a century ago railway engineers discovered that to make maximum use of power developed, maximum adhesion was essential, so they abandoned the great single driving wheels and coupled four, six and even ten wheels together to get adhesion. It is just the same with a road vehicle. If you want maximum adhesion under bad conditions, two driving wheels are not enough. Four-wheel drive is the answer.

We proved that with the Land Rover on deep snow and iced roads. At 1,800 feet it was driven in a blinding snowstorm and a gale of 50 m.p.h., along glassy roads with a drop of 100 feet to one side, and it never wavered. With four-wheel drive engaged it safely overtook faster cars on the frozen hilly roads of Herefordshire and the snow-covered Cotswolds.

A lot of good hard commonsense has been put into the design, and, after the exasperating effeminate fripperies that adorn some modern cars it was a joy to find anything so solid, sound and to hand. There were no control knobs that came off in your hand, no plastic ashtrays that broke when you hit 'em with a pipe, no tiny control pedals designed for a Geisha girl's foot. And, of course, there was nothing really

(Above) Capacity of the standard body is shown in this picture with the tilt removed. The permitted load is 1,000 lb. (plus three passengers), and the spare wheel can be mounted on the bonnet if necessary. The maximum drawbar pull is put at 2,000 lb., but in practice it will move more than that from a standstill. (Left) Trailer and straw together weigh about four tons, but the Land Rover had no difficulty whatever in hauling it. The farm and field jobs possible with this highly versatile vehicle are so considerable that it is hoped, at a later date, to try out a further model on these jobs.

urious about it. The car is
ctly functional.

A single outside mirror is fitted
d serves well. But there is a
nounced blind spot (with the
up) on the nearside. This
ld be cured by a wide window
the tilt, but such a safety
vision would bring down a
cking burden of purchase tax.
e remedy seems to be another
ror on the nearside wing.

The whole car is built to take
d knocks. Instead of chromium
re is galvanised steel; instead of
lded door panels, there is non-
roding aluminium; instead of
pely meaningless bumpers that
't take a bump, there's a great
vanised steel fender that could-
sh a steam roller.

Springing is firm, and is
pported by telescopic hydraulic
ck absorbers. The springs are
in. wide, 36½ in. long on the front
l 48 in. on the rear. To provide
ra strength for cross-country
rk, the second leaves are curled
und the shackle pin eyes.

The Land Rover is primarily
al; there's nothing urban about
This characteristic extends to
h things as the petrol filler cap
the 10-gallon tank under the
ver's seat. This cap is so big that
rol could be poured from a
cket into it, without splashing
sides. The filler incorporates a
ge filter which is always highly

desirable especially when there is
much filling from cans.

Brake and clutch pedals, covered
with deep ribbed rubber pads, are
large enough for a driver wearing
size 13 gumboots. The accelerator
pedal is a piece of flat steel, so long
and wide that no driver's foot could
miss it.

The unit, of 1,997 c.c., has overhead inlet and side exhaust valves.
It is long wearing, extremely
efficient, and its note tells of the
power there. Compression ratio is
6.9 to 1, and while it behaved well
on commercial grade fuel, the best
results were on a 50-50 mixture of
commercial and premium grades.
Maximum output is 52 b.h.p. at
4,000 r.p.m.

Since the Land Rover is intended
for a good deal of stationary work,
cooling has to be adequate. This is
looked after by a radiator holding
17 pts., and a four-bladed fan
running close cowled.

This enables engine to run all day
stationary without over-heating,
but in the Arctic conditions of the
test there was difficulty in getting
the engine to its best running
temperature in road work. This
could probably be overcome by
fitting one of the radiator blinds
now on the market—or, as we did,
simply by covering up part of the
radiator. An engine temperature
gauge would be a worthwhile
addition to the instruments.

In normal road work, of course,
four-wheel drive is not used. While
it is automatically in operation
when the low ratio gearbox is used,
it is optional on the high ratio box,
and is brought in by depressing a
central lever. Both gearboxes offer
four speeds and reverse and top and
third on the main gearbox are
synchromesh. This gearbox was
delightful to use.

The fact is that it required really
exceptional gradients to make the
use of the low transfer box necessary.
It was used under the worst off-the-
road conditions in Wales, when
negotiating deep fords (after which
the brakes froze momentarily), and
when climbing (in reverse) a bank
on the alpine course so steep that
the test team climbed up it on all
fours to take photographs.

It was a testimony to the remarkable strength of the hand brake
acting on the transmission, that it
held the car on this bank. This
brake is really for parking and is not
intended for use while moving. But
in traffic, as for instance when
holding the car on a slope at traffic
lights, it sometimes has to be used
for re-starting, and then the position of the lever, almost on the floor,
becomes a trifle inconvenient. There
is a minimum of linkage on this
control and its position is out of the
way of the central passenger's foot.
All the same, if the lever could be
cranked, or otherwise raised, it
would be far more convenient for
the driver's hand.

The foot brakes are hydraulic
and exceptionally good. On test
they registered (dry roads) 76 per
cent from 20 m.p.h., and the very
good figure of 94 per cent from 30
m.p.h. (100 per cent = 30 ft. from
30 m.p.h.).

On the weighbridge, the Land
Rover scaled, in running trim with
only the driver up, 28 cwt, applied
as 15 cwt on the front axle and
13 cwt on the rear.

Maximum payload is around
1,000 lb. and passenger capacity
(on this general purpose model)
seven.

In the coldest conditions we
found the steering inclined to be
on the heavy side. It is always
firm and precise—the car goes
where you point it, and naturally,
with a driven front axle it is
inevitably heavier.

A sensible provision is an amber
warning light which goes on when

the engine temperature is sufficient
to dispense with the choke. With
commercial fuel on very cold mornings a good deal of choke was
necessary.

Undoubtedly this and the difficulty of reaching a high enough
engine temperature during road
work in these bitter conditions,
affected the petrol consumption.
In these circumstances the figure of
22.56 m.p.g. for the first 677 miles,
including a good deal of off-the-
road low ratio work, and the test
on the alpine course, was not un-
satisfactory. After a spate of
towing heavy loads, a second spell
of alpine work and some driving in
London traffic, consumption
dropped to 21.45 m.p.g. for 1,000
miles. All the same, the maker's
claim of 23 m.p.g. on road work
under normal conditions, is probably quite easy to attain. Acceleration was good. The figures were:—
0-30 m.p.h. (starting in first and
through the gears)—10 secs.; 0-30
(starting in second—mean of three
runs)—8.5 secs.; 0-40 m.p.h. (starting in second)—14.25 secs.; 0-50
m.p.h.—22.5 secs. Maximum road
speed was about 60 m.p.h., and the
car cruised comfortably at 45 m.p.h.

The model tested was fitted with
an eight-hole drawbar plate and
large and small towing jaws which
were bolted on in a few minutes.
There is also an extension plate for
the towing jaw so that this can be
used while the rear power take-off
is still in position. Besides rear
p.t.o., there is also a central p.t.o.
(both at extra cost) which can be
fitted with pulleys of various types.
A capstan winch can be fitted to
the front for recovery and haulage
work.

So far as farmers are concerned,
the standard model tested, even
without the various attachments, is
probably the most useful British
vehicle ever built—either as a
transporter of small livestock, supplies or farm staff, a puller of trailers
of all sorts and sizes, a general
tender, or a go-anywhere runabout,
something that stands around the
farm thick with mud, gets—and
needs—little attention yet cannot
rust and always goes well just when
and where you want it.

At £585 (there is no purchase
tax on the standard model) it is
for the farmer probably the best
universal transport value for money
available today.

ove) Note the axle
itions in this picture.
s, even when frozen
d, had no effect on
e rigid chassis.
ht) Inside the bonnet
battery (left), oil-
h air cleaner, car-
rettor and dis-
utor are easily ac-
sible. (Right, centre)
iver's and pas-
gers' seats are
wn as well as the
n-sized control
als. (Far right) The
rol tank below the
ver's seat has a filler
p (incorporating
er). Note the hand-
ke lever and the link
matting fitted.

The Motor Road Test No. 18/56

Make: Rover **Type:** Land Rover "107" Station Wagon de Luxe
Makers: The Rover Co. Ltd., Solihull, Birmingham.

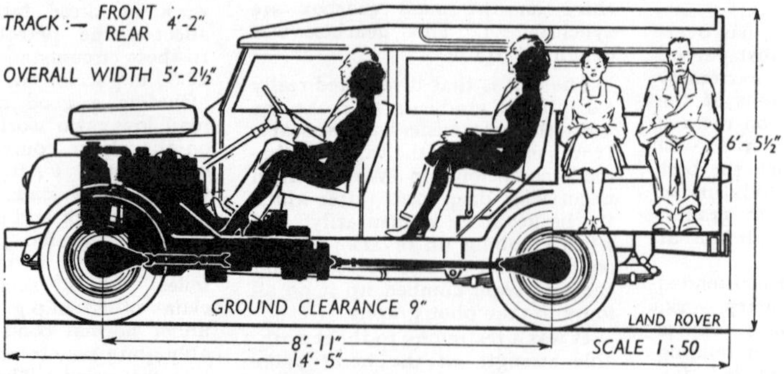

TRACK:— FRONT / REAR 4'-2"
OVERALL WIDTH 5'-2½"
6'-5½"
GROUND CLEARANCE 9"
LAND ROVER
8'-11"
14'-5"
SCALE 1 : 50

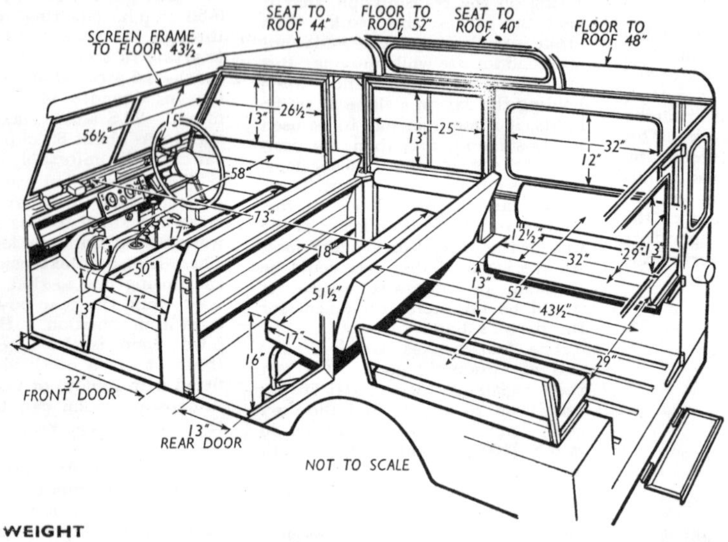

SCREEN FRAME TO FLOOR 43½"
SEAT TO ROOF 44"
FLOOR TO ROOF 52"
SEAT TO ROOF 40"
FLOOR TO ROOF 48"
56½"
15"
13"
26½"
25"
13"
32"
12"
58"
73"
17"
12½"
32"
29"
13"
18"
50"
17"
13"
13"
52"
51½"
43½"
17"
16"
29"
32" FRONT DOOR
13" REAR DOOR
NOT TO SCALE

Test Data

CONDITIONS. *Mild, showery weather with gusty wind (dry for brake test). Temperature 59-62°F., barometer 30.0 in. Hg. Smooth tarred road surface. Standard-grade pump fuel.*

INSTRUMENTS

Speedometer at 30 m.p.h.	4% slow
Speedometer at 60 m.p.h.	2% slow
Distance recorder	3% slow

MAXIMUM SPEEDS

Flying Quarter Mile

Mean of four opposite runs	58.1 m.p.h.
Best time equals	60.4 m.p.h.

"Maximile" Speed (Timed ¼-mile after one mile accelerating from rest)

Mean of four opposite runs	57.2 m.p.h.
Best time equals	58.8 m.p.h.

Speed in gears

Maximum speed in 3rd gear	51 m.p.h.
Maximum speed in 2nd gear	37 m.p.h.
Maximum speed in low top gear ..	31 m.p.h.
Maximum speed in low 3rd gear ..	22 m.p.h.

FUEL CONSUMPTION

26½ m.p.g. at constant 30 m.p.h.
23 m.p.g. at constant 40 m.p.h.
19½ m.p.g. at constant 50 m.p.h.
Overall consumption for 998 miles, 55 gallons =18.2 m.p.g. (15.5 litres/100 km.)
Fuel tank capacity 10 gallons.

ACCELERATION TIMES Through Gears

0-30 m.p.h.	7.8 sec.
0-40 m.p.h.	14.7 sec.
0-50 m.p.h.	28.9 sec.
Standing Quarter Mile	26.2 sec.

ACCELERATION TIMES on Two Upper Ratios

	Top	3rd
10-30 m.p.h.	12.7 sec.	8.2 sec.
20-40 m.p.h.	15.7 sec.	11.9 sec.
30-50 m.p.h.	24.7 sec.	—

WEIGHT

Unladen kerb weight	30¾ cwt.
Front/rear weight distribution ..	51½/48½
Weight laden as tested	34¼ cwt.

HILL CLIMBING (At steady speeds)

Max. gradient on top gear	1 in 12.7 (Tapley 175 lb./ton)
Max. gradient on 3rd gear	1 in 8.7 (Tapley 255 lb./ton)
Max. gradient on 2nd gear	1 in 6.7 (Tapley 330 lb./ton)

BRAKES at 30 m.p.h.

0.93g retardation ..	(=32½ ft. stopping distance) with 140 lb. pedal pressure
0.67g retardation ..	(=45 ft. stopping distance) with 100 lb. pedal pressure
0.32g retardation ..	(=94 ft. stopping distance) with 50 lb. pedal pressure
0.06g retardation ..	(=500 ft. stopping distance) with 25 lb. pedal pressure

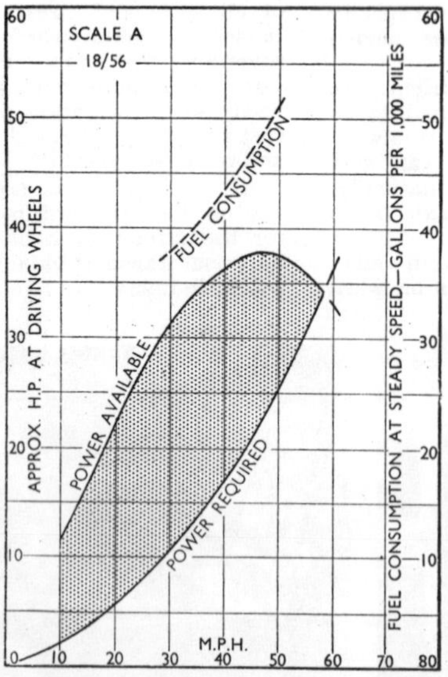

SCALE A
18/56
FUEL CONSUMPTION
POWER AVAILABLE
POWER REQUIRED
APPROX. H.P. AT DRIVING WHEELS
FUEL CONSUMPTION AT STEADY SPEED—GALLONS PER 1,000 MILES
M.P.H.
0 10 20 30 40 50 60 70 80
10 20 30 40 50 60

Drag at 10 m.p.h. .. .. .. 80 lb.
Drag at 60 m.p.h., by extrapolation approx. 227 lb.
Specific Fuel Consumption when cruising at 80% of maximum speed (i.e. 46.5 m.p.h.) on level road, based on power delivered to rear wheels .. .. 0.80 pints/b.h.p./hr.

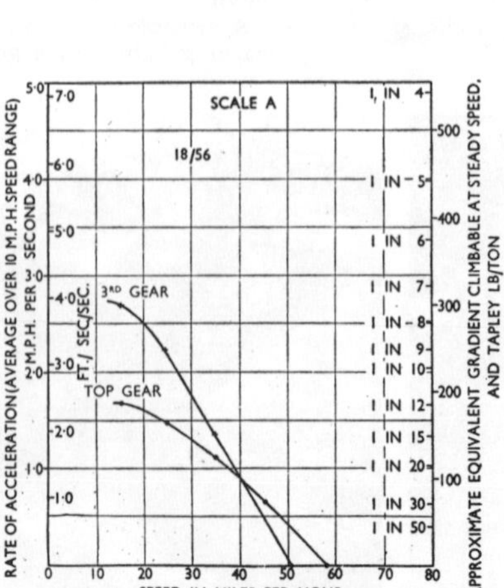

SCALE A
18/56
3RD GEAR
TOP GEAR
RATE OF ACCELERATION (AVERAGE OVER 10 M.P.H. SPEED RANGE) M.P.H. PER SECOND
FT./SEC./SEC.
SPEED IN MILES PER HOUR
APPROXIMATE EQUIVALENT GRADIENT CLIMBABLE AT STEADY SPEED. AND TAPLEY LB/TON
1 IN 4
1 IN 5
1 IN 6
1 IN 7
1 IN 8
1 IN 9
1 IN 10
1 IN 12
1 IN 15
1 IN 20
1 IN 30
1 IN 50
500
400
300
200
100
0 10 20 30 40 50 60 70 80

Maintenance

Sump: 10 pints, S.A.E. 30 summer, S.A.E. 20 winter. **Gearbox:** 2½ pints, S.A.E. 90 gear oil. **Transfer Gearbox:** 4½ pints, S.A.E. 90 gear oil. **Front & Rear Axles:** 3 pints each, S.A.E. 90 E.P. gear oil. **Swivel pin housings,** 1 pint each, S.A.E. 90 E.P. gear oil. **Steering gear:** S.A.E. 140 gear oil. **Radiator:** 17 pints (2 drain taps). **Chassis Lubrication:** By oil gun every 3,000 miles to 6 points. **Ignition timing:** 10° B.T.D.C. static. **Spark Plug gap:** 0.029 — 0.032 in. **Contact breaker gap:** 0.014—0.016 in. **Valve timing:** No. 1 Exhaust valve peak 114° B.T.D.C. **Tappet clearances** (hot or cold) Inlet 0.010 in., exhaust 0.012 in. **Front wheel toe-in:** 3/64 in. to 3/32 in. **Camber angle:** 1½°. **Castor angle:** 3°. **Tyre pressures:** Front 25 lb., Rear 25 lb. (for loads exceeding 550 lb. raise rear pressure to 32 lb., for very soft ground lower tyre pressures to 16 lb. minimum unladen, 24 lb. minimum fully laden.) **Brake fluid:** Girling Crimson. **Battery:** 12-volt 51 amp hr.

Ref. B/20/56

The Land Rover "107" Station Wagon

A Rugged Ten-seat "Go-anywhere" Vehicle which is also Pleasant to Drive on the Road.

AT HOME on the road or off it, the long-wheelbase station wagon bears out the words of the Balkan mining engineer who said: "With my Land Rover I go where only asses go"! In the situation on the left, however, it was held on the handbrake and re-started comfortably, scraping its armoured belly on the crest but climbing over without trouble.

VERSATILITY, sturdiness and a total lack of decoration are the features which make the Land Rover station wagon unique amongst the products of the British motor industry. It is easy to drive, tolerably comfortable and capable of 60 m.p.h. in slightly favourable conditions, so that it can do the work of a normal car. There are seats for ten people including the driver, enabling it to do the work of a small bus. Seven of the seats can be folded instantly out of the way, or removed completely in a very few minutes, leaving the roominess of a light van approved for carrying loads of more than half a ton. Finally, there is four-wheel-drive and a special set of low gears to be engaged if desired, these and a very generous ground clearance ensuring mobility on atrociously bad roads or right away from roads.

Inevitably, certain limitations go with such exceptional versatility. Performance figures are best compared with those for an 8 h.p. car of orthodox design, with substantially more acceleration from low speeds in top gear but a slightly lower maximum speed. Fuel consumption is heavier than with the same 2-litre engine installed in an orthodox car, owing to the weight and wind resistance of this roomy vehicle. Springing does not provide the smoothness of modern cars designed to carry less widely varying loads over sur-

faced roads. The first cost inevitably reflects the extra complication of four-wheel-drive and the careful engineering which, for example, virtually eliminates servicing for 3,000 mile intervals. But for undeveloped parts of the world and for some arduous tasks in Britain, too, the Land Rover Station Wagon has tremendous advantages.

When the Land Rover was introduced in April, 1948, it was as an open-bodied vehicle with 80-inch wheelbase and 1.6-litre engine. Since that time, the engine has been enlarged to 2 litres swept volume, the normal wheelbase increased to 86 inches, and an alternative chassis of 107-inch wheelbase added to the range. The latter is the subject of this test report, with the station wagon type of body which is offered as alternative to an open truck with enclosed 3-seat cab. In de-luxe trim, with rubber floor covering, washable plastics upholstery, interior lighting and heating, windscreen de-misters, etc., this rugged vehicle was used as an ordinary car around London and the Home Counties in addition to taking it into tougher surroundings.

Slightly shorter in overall length than the ordinary 6-seat Rover car, this 10-seat model will fit into normal garage and car spaces so long as its overall height of 6 ft. 6 in. can be accommodated. Ground clearance which is 8 inches at the minimum and generally much greater, is re-

flected in a rather high floor level, which in conjunction with upright seats makes the Land Rover higher than a modern car for entry or exit. Once entered, however, it is roomy, with upright and non-adjustable but well cushioned seats and a lot of headroom. Although a tall driver cannot stretch his legs, quite long spells at the wheel can be tackled before a rest seems desirable.

Ignoring two extra controls which need not be touched when on surfaced roads, the Land Rover is just like an ordinary car to drive, with a central lever controlling a four-speed gearbox which has synchromesh on the upper two ratios. The commanding outlook, across the bonnet upon which the spare wheel is mounted, and over the roofs of many other cars, and the huge brake and clutch pedals, are suggestive of a lorry, but the controls are as smooth as on a car and lighter and more positive than on some more ordinary vehicles.

Although a choke is provided, it was not needed during our test, settings in use on the pump-type Solex carburetter being such that slight "jiggling" of the accelerator sufficed for starts from cold after summer nights of parking out of doors, the tick-over immediately becoming perfectly reliable. Most of our testing

In Brief

Price: £790 plus purchase tax £396 7s. 0d. equals £1,186 7s. 0d.

Capacity	1,997 c.c.
Unladen kerb weight ...	30¾ cwt.
Fuel consumption ...	18.2 m.p.g.
Maximum speed ...	58.1 m.p.h.
"Maximile" speed ...	57.2 m.p.h.
Maximum top gear gradient	1 in 12.7

Acceleration:
10-30 m.p.h. in top ... 12.7 sec.
0-50 m.p.h. through gears 28.9 sec.

Gearing: 15.95 m.p.h. in top at 1,000 r.p.m.; 57.9 m.p.h. at 2,500 ft. per min. piston speed.

THREE compartments provide accommodation for ten people in the long-wheelbase station wagon, the two forward sections (*below*) providing comfortable three-abreast seating once the high step-up from ground level has been made. In these pictures, the double-skin roof and the sky-light over the side door are clearly visible.

The Land Rover "107" Static

hustling driver is more than able to hold his own with other traffic in this vehicle.

The extra mechanism of four-wheel drive has not made the steering either heavy or unresponsive for brisk driving, although when manoeuvring at the lowest speeds rather more work has to be done on the wheel. Despite some friction which eliminates kick-back, the steering allows the car to be placed accurately by a driver who is familiar with the vehicle, and the absence of need for lubrication other than the occasional checking of oil reservoir levels suggests that handling characteristics should deteriorate very little with the passing of time.

Stiff But Steady

Decidedly firm springs are used on this vehicle, much firmer than on modern cars although not so harsh as on the four-wheel-drive vehicles which many motorists rode in during the war. Extra wheelbase length has certainly given this vehicle better riding, but with high unsprung weight and firm springs comfort depends upon speed being moderated over really bad bumps. As a reward for firmness, there is a pleasant freedom from roll or sway during brisk cornering even with the vehicle well laden. The brakes need rather higher pedal pressures than are nowadays usual on cars, but are not lacking in power and survived brisk negotiation of shallow fords without disappearing altogether. Although its lever is rather awkwardly placed right below the driving seat, the transmission type handbrake proved quite able to hold this heavy vehicle on slopes steeper than those on to which any ordinary car would venture.

On ordinary roads, 50 m.p.h. is a perfectly happy cruising speed, with another 5 m.p.h. fairly readily available, but fast cruising is hard on the fuel consumption and rather noisy—especially if the excellent ventilation flaps below the windscreen are opened in hot weather. At all times, the transfer gears needed with four-wheel drive are audible to a certain extent. Off the road, an auxiliary gear lever with a red knob may be moved back, simultaneously to link the drive to the front wheels and to bring an "underdrive" of approximately 2½-1 ratio into use between the ordinary gearbox and the propeller shafts.

In the low range of gears, one valuable ability of the Land Rover is to travel really slowly, for example when pushing aside

INWARD-FACING seats are used in the rearmost compartment which is entered by a wide back door with a folding step. For use as a van, seats may be folded away or completely removed, or they may be re-arranged to form a large double bed, as seen below.

was carried out on commercial-grade fuel, with which in use only very slight pinking at full throttle and low speeds was evident. The engine was not completely smooth below 20 m.p.h. in top gear, but in general proved itself extremely docile for a four-cylinder design, and reasonably quiet despite the lack of any sound deadening treatment upon the body's many flat panels.

Ordinary car ratios are used in the gearbox, which is fairly quiet and can be advantageously used for overtaking other main-road traffic up to over 40 m.p.h. No synchromesh is provided on 2nd gear, but it is an easy downward change from 3rd, and the upward change from 1st need seldom be made unless heavy loads are being started from rest uphill. A figure of 12.7 seconds for top gear acceleration from 10 to 30 m.p.h. indicates liveliness comparable with many 1-1½-litre cars, the acceleration diminishing gradually towards a two-way mean maximum speed of 58.1 m.p.h. On busy suburban roads, a

STRENGTH and simplicity are evident in the driving compartment. Pedals suitable for heavy boots are not uncomfortable for soft-shod drivers, and the steering is surprisingly light. Alongside the central gear lever are levers for engagement of four-wheel drive and for introduction of low-ratio gears.

Wagon

the branches of trees in forcing its way through woods or down overgrown tracks, or when getting into sharp hollows or over humps. It will also of course climb any gradient upon which four bodly-treaded tyres can find grip, low top gear providing a maximum of 30 m.p.h. although it is best to engage a gear in high range if more than 20 m.p.h. is being sustained. Provision for slippery going which is not otherwise very difficult takes the form of a yellow knob which, when pressed down, engages four-wheel drive without use of the low range of gears.

Our test was made in British summer weather when deep mud and slippery surfaces were difficult to find, but obviously the Land Rover's ability to cover rough ground is little impaired by its lengthened wheelbase. One weakness of this vehicle, however, is the far-from-compact turning circle, the shorter alternative chassis being obviously better in this respect.

Our test model came equipped for either hot or cold weather, a double skin roof with four opening ventilators keeping it cool by day, and a re-circulating heater warming the front compartment only at night.

An excellent interior light is provided, all doors can be locked as can the sliding side windows, and two separate motors drive the windscreen wiper blades at speeds which differed by an irritating 5%. Lifting the bonnet with the spare wheel in place is emphatically a man's job, but the over-centre stay provided to keep it up seems safe, and although there is no waste space around the engine accessibility for routine jobs is good.

Apart from the choice of open or closed bodywork on a long or a short chassis, the Land Rover can be equipped to do many and varied jobs. As well as being offered as a go-anywhere light fire engine, it can be equipped with a winch on the front bumper for timber hauling; with power take-offs on the back bumper to drive farm machinery by shaft or by flat belt; with a power take-off amidships to drive fixed equipment by triple vee-belt; or, of course, with fittings for all kinds of towing, be it for a holidaymaker's caravan or a farmer's harrow. It is noteworthy that the Land Rover body is rust proof, built of incorrodible aluminium alloy panels and galvanized steel framing.

Capacious, seemingly of almost unbreakable strength, and certainly no sluggard, the Land Rover used (cheap) fuel fairly quickly, steady speed recordings varying from 26½ m.p.g. at 30 m.p.h. down to 19½ m.p.g. at 50 m.p.h. on smooth and level road. Our overall figure of 18.2 m.p.g. for 998 miles includes as usual more fast driving than slow, city traffic and many starts from cold as well as exploration of places which ordinary cars cannot approach. Using more fuel than would an ordinary car of medium size, the Land Rover is in fact probably more economical than any comparable vehicle produced elsewhere in the world, despite which its fuel capacity of 10 gallons (in a tank below the driving seat which can easily be filled from cans) seems disappointingly small.

Most buyers of the Land Rover Station Wagon probably do not choose it—there is nothing else available which will do so much in such difficult conditions, so they buy it almost inevitably. The facts that it gives promise of exceptional durability, and can be enjoyed as a car to drive or ride in, are just so much bonus value.

Mechanical Specification

Engine

Cylinders	4
Bore	77.8 mm.
Stroke	105 mm.
Cubic capacity	1,997 c.c.
Piston area	29.5 sq. in.
Valves	Pushrod o.h. inlet, side exhaust
Compression ratio	6.7/1 (optional 6.9/1)
Max. power	52 b.h.p.
at	4,000 r.p.m.
Piston speed at max. b.h.p.	2,760 ft. per min.
Carburetter	Solex 32 PBI-2 downdraught
Ignition	12-volt coil
Sparking plugs	14 mm. Lodge CLN-H
Fuel pump	S.U. electrical
Oil filter	AC-Delco full-flow

Transmission (4-wheel drive)

Clutch	9-in. single dry plate
Top gear (s/m)	5.396 (low range, 13.578)
3rd gear (s/m)	7.435 (low range, 18.707)
2nd gear	11.026 (low range, 27.742)
1st gear	16.171 (low range, 40.688)
Propeller shafts	Hardy Spicer, open
Final drive	4.7/1 spiral bevel
Top gear m.p.h. at 1,000 r.p.m.	16.0 (low range, 6.35)
Top gear m.p.h. at 1,000 ft./min. piston speed	23.2 (low range, 9.2)

Chassis

Brakes	Girling hydraulic, 2 l.s. front (separate parking brake on transmission)
Brake drum diameter	11 in.
Friction lining area	133.8 sq. in.
Suspension:	
Front	Semi-elliptic
Rear	Semi-elliptic
Shock absorbers	Monro-matic telescopic
Tyres	7.00—16

Steering

Steering gear	Burman re-circulating ball
Turning circle (between kerbs):	
Left	42 feet
Right	46 feet
Turns of steering wheel, lock to lock	4¼

Performance factors (at laden weight as tested)

Piston area, sq. in. per ton	17.22
Brake lining area, sq. in. per ton	78
Specific displacement, litres per ton mile	2,190

Coachwork and Equipment

Bumper height with car unladen:	
Front (max.)	22 in., (min.) 19 in.
Starting handle	Yes
Battery mounting	Alongside engine
Jack	Screw type
Jacking points	Under axles
Standard tool kit:	Tyre pump and gauge, wheelbrace, adjustable spanner, single-ended spanner, 2 double-ended spanners, sparking plug spanner and tommy bar, pliers, screwdriver, distributor screwdriver/feeler, oil gun.
Exterior lights:	2 headlamps, 2 sidelamps/flashers, 2 stop/tail/flasher lamps.
Direction indicators	Flashers, non self-cancelling
Windscreen wipers:	2 separate single-blade electric wipers, non self-parking
Sun vizors	None
Instruments:	Speedometer with non-decimal non-trip distance recorder, ammeter, fuel content gauge.
Warning lights:	Dynamo charge, oil pressure, rich mixture control, headlamp main beam
Locks:	
With ignition key	Ignition
With other key	Driver's and rear doors
Glove lockers	Open lockers on each side of facia panel
Map pockets	2 in front doors
Parcel shelves	None
Ashtrays	None
Cigar lighters	None
Interior lights	One in roof
Interior heater	Optional extra, re-circulating type with screen de-misters
Car radio	Optional extra
Extras available:	Chaff guard, universal joint covers, heavy duty towing pintles, rubber pads for pedals, radio interference suppressors, combined water thermometer and oil pressure gauge, radio installation, oil cooler equipment, road speed limiting governor, car heater and de-mister, hand throttle control, flyscreens for dash ventilators, front lifting and towing rings, front capstan winch.
Upholstery material	Vynide plastics
Floor covering	Rubber
Exterior colours standardized	Grey or blue
Alternative body styles:	Open pick-up truck (also short-chassis models)

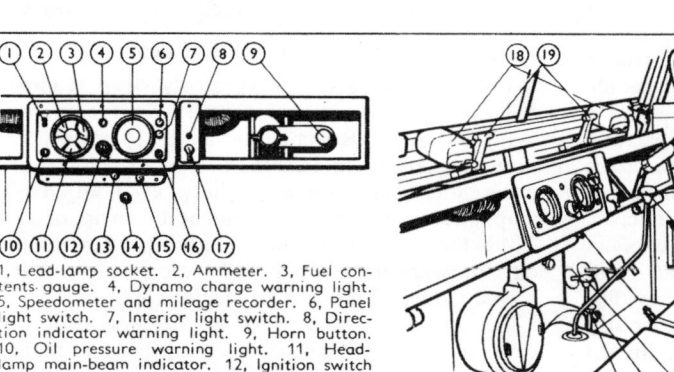

1, Lead-lamp socket. 2, Ammeter. 3, Fuel contents gauge. 4, Dynamo charge warning light. 5, Speedometer and mileage recorder. 6, Panel light switch. 7, Interior light switch. 8, Direction indicator warning light. 9, Horn button. 10, Oil pressure warning light. 11, Headlamp main-beam indicator. 12, Ignition switch and head and side-lamp switch. 13, Heater fan switch. 14, Starter button. 15, Choke control. 16, Mixture control warning light (choke closed). 17, Direction indicator switch. 18, Windscreen wiper motors and controls. 19, Air vent operating levers. 20, Gear lever. 21, Handbrake. 22, Low-ratio gear transfer. 23, Headlamp dip-switch. 24, Front-wheel drive engagement control (push down to engage).

THE ROAD FROM BETHLEHEM.—
A magnificent panorama across
the Vale of Towy unveiled itself
on the run down from the Welsh
Bethlehem towards Llandeilo. On
the right is the Land Rover/Eccles
E.16 outfit in which the journeys
were made.

MYNYDD EXCURSION

Land Roving with an Eccles in Wales' not so "Wild and Woolly" West, with some Motor Mountaineering Interludes

QUITE a lot of people use Land Rovers for hauling caravans. Until recently, however, all my towing has been done with ordinary motorcars and, in particular with a Triumph Roadster which has now served for 84,000 miles. Thus, when the arrival of a Land Rover Station Wagon on the 107-in. wheelbase, for *The Motor* technical department to Road Test, coincided with a suggestion from the Eccles concern that I should do a user test on a prototype of their recently introduced E.16 model, the tie-up was obvious and the loan of the Rover extended.

Almost immediately, one great advantage of such a vehicle from a town-dwelling caravan user's point of view became apparent. In the normal way, a caravan has to be loaded up at home which usually presents parking problems. With the Station Wagon, the 'van could be left out in the country on our projected route, in this case at a little inn in Crowell on the northern slopes of the Chilterns, and the whole party (four people and a Sealyham) together with all requirements for a fortnight, ferried out to it on the eve of departure. Thus, on the following morning we were 36 miles towards our objective, unruffled by the frustrations of an early morning city exit against the incoming traffic.

It needed but a few miles with the fully-laden caravan in tow to discount rumours I had heard that Land Rovers are rather easily influenced by caravan behaviour. Seldom have I driven a more stable car/caravan outfit despite the fact that the latter was not specially loaded to provide the front-end heaviness considered essential to ensure good following, and was riding slightly nose high which is generally regarded as *not a good thing*. Mark you, the Eccles E.16 is an inherently stable caravan; it was, however, possible purposely to induce

snake, and at such times the behaviour of the trailing vehicle could be seen but not felt.

People who are in the habit of making high-speed, caravan-laden dashes to the South of France may find the Land Rover's rather low maximum speed a bit frustrating for, with a ton and a bit of 'van in tow, 48-50 m.p.h. seemed to be its absolute maximum no matter how long a stretch was available to build up pace. On this side of the Channel though, indeed anywhere where the roads wind and gradients are frequent, there is ample speed and, with intelligent use of the easy-to-change ratios, all the accelerative performance needed to maintain a good cruising pace. In this regard, the fact that one sits high ensures ample see-through in the rear mirror at what is going on behind, in addition to providing a good anticipatory forward view.

No Fear of Gradients

Fear of failure on unexpectedly steep gradients frequently keeps many caravanners on the main roads. With the Land Rover's low auxiliary ratios, giving a bottom gear of no less than 40.68 : 1, plus four-wheel drive, such fears are groundless: even an involuntary stop-and-restart on 1 in 4 can be taken in one's stride. Thus, there are plenty of opportunities for taking short cuts or wandering along little-used roads to get better scenic value. Take our particular case.

I have long used A40 to Pembrokeshire with a caravan in tow, always sticking to the main route although I loath the winding, narrow section between Brecon and Llandovery, particularly since the occasion when, coming up on the tail of a slow-moving army convoy, it took the best part of two hours to cover the 21 miles. There is, however, a mountain road running

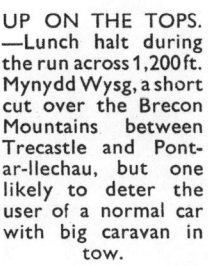
UP ON THE TOPS. —Lunch halt during the run across 1,200 ft. Mynydd Wysg, a short cut over the Brecon Mountains between Trecastle and Pont-ar-llechau, but one likely to deter the user of a normal car with big caravan in tow.

MYNYDD EXCURSION

from Trecastle to Llandeilo which cuts off a respectable mileage and by-passes Llandovery altogether. Rising at the start 700 feet in little over a mile, it crosses the south-eastern escarpment of Mynydd-bach Trecastell (Little Trecastle mountain) and thence goes slap across Mynydd Wysg (pronounced "ooisg") joining the Swansea—Llangadock road for a couple of miles and then diving off through Bethlehem to Llandeilo. Despite a total power to weight ratio of only 17 b.h.p./ton, this time we shot off over the short cut without a qualm, thereby not only saving mileage but being rewarded, although there was low cloud, with some impressive upland scenery and magnificent views across the lower stretches of the Towy valley.

Threat of Popularity

Ultimate destination was Western Pembrokeshire where, for many past years, the hurly-burly of London has been exchanged for the peace of a little known coast, whenever opportunity afforded, although the weather is problematical and conforms to no forecast unless it be that relating to shipping in the Lundy area. It used to be one of the few areas within reasonable reach of the metropolis where one could really be "out in the sticks," far from the tripper, the ice-cream parlour and the dead hand of officialdom. But that is all being changed. Frightened no doubt by the influx of visitors during last year's unusually fine summer, and aided and abetted by the B.B.C., who ran a feature programme on the beauties of the Pembrokeshire coast (from a tourist angle), the local county council are "doing something about it" in the way of lay-by's on peaceful country lanes, car parks where one used to just drive on to the grass, and "No parking" notices elsewhere to ensure that the parks are well patronized. In short, the area is being "popularized" and, at the rate things are going may, like Devon and Cornwall. soon become a trippers' paradise with all its attendant evils.

PEACEFUL PITCH. — Base of operations was this secluded spot at the head of a small Pembrokeshire haven. Despite its built-in aerial, the portable radio needed jury-rigged, elevated assistance to ensure reception in this "semi-dead" area.

LIST TO PORT.—Even the Land Rover had to be driven at some peculiar angles on the climb up Mynydd Prescelly.

MYNYDD EXCURSION

Highest point of the area on the eastern edge of the country is Mynydd Prescelly, dominated by 1,760-ft. Prescelly Top. From its summit there is a magnificent view reaching away, on a clear day, to Ireland, Devonshire and Carnarvonshire. The temporary possession of a vehicle with "off-the-road" abilities suggested that the ascent might be made without recourse to pedestrian antics. With a normal car it would, of course, be quite out of the question; in fact, I doubt whether one would get much more than

SUCCESS AND FAILURE.—First attempt at 1,760-ft. Prescelly Top was crowned with success, as the photograph (above) by the Bench Mark indicates. A second ascent, after 24 hours of thunderstorms, resulted in the Land Rover becoming bogged even beyond the scope of its four-wheel drive and low auxiliary ratios, resulting in a six-hour wait for tractor-assisted extraction.

100 yards across the lower slopes of the mountain.

"Watch out for the reeds," they warned us when we discussed the project in the Tafarn Newydd (New Inn, to you) at Rosebush—a nice little pub whose stark exterior belies its inner comfort. So, unaccompanied by the ladies of the party, we rocked, bumped, swayed and spun our way upwards to where the reeds begin, some 500 ft. below the summit. Lots of reconnaissance showed a way across. A flat-out charge in low third and we were through the mire with but a short, if bumpy, climb remaining to the top.

Two days later, the womenfolk resolving the view to be worth a possible shaking up, we set off again with a full complement of passengers, to discover that, excellent vehicles though they may be, Land Rovers are lacking in one particular—they can't swim. Once before this omission has been demonstrated—again in a Welsh mountain bog—and called for pedestrian navigation of Rhayader's biggest reservoir in search of help. This time, although the wheeltracks of our previous ascent were discovered and followed, allowance had not been made for the amount of water that can accrue from 24 hours of thunderstorms. Down we went to the chassis, through the thick moss, to the water below.

In these parts, farmers have to take the fullest advantage of every fine day to gather in their winter fodder, and six hours were spent on the windswept height while the nearest tractor-owning farmer finished his haymaking before coming to our aid.

Despite the fact that our Pembrokeshire stay was punctuated by many excursions over tracks, generally marked "Unsuitable for Motors," but leading to attractive, unfrequented parts of the coast, that was our only

involuntary halt. But it must be recorded that many voluntary halts were made on corners of steeper radius than the Rover's rather inadequate steering lock.

From start to finish, careful check was kept of the amount of fuel used, both with the caravan in tow and without it. Of the 508 towing miles, 14.7 were covered for each gallon of standard-grade fuel, as against 19.8 miles on a similar quantity when motoring unencumbered. Thus, our journeyings with 21 cwt. of caravan behind cost 27s. 9d. per 100 miles—which, considering the hilly nature of much of the course, we considered reasonable enough.

Apart from the fact that it towed well, little has been said about the caravan, which was our base of operations during the period. Like most of the products of the Eccles concern, the E.16 is an essentially practical vehicle, but with a difference. While the general layout—two singles and a double berth, end kitchen and toilet room—is fairly orthodox, there are a number of features, including an ingenious sliding partition which does not involve using the wardrobe door as part, a proper kitchen cabinet with drop-down working surface, and quite an amount of free floor space and cupboard room, which are desirable, but not always obtainable, in a 'van of such size (16 ft.) and price (£410). Added to which, the makers have obviously paid a lot of attention to "eye-appeal" and the inside is positively gay. The new lines, too, are a departure from orthodox Eccles, but none the less attractive for that.

As was to be expected in a prototype which was only finished shortly before it left the works behind the Land Rover, a few "sillies" were discovered by actual use, but these have since been rectified in production models: they were all of a minor nature and had little effect on our general comfort, parked as we were at the head of a quiet little haven with ev. mod. con. at our disposal. E.H.R.

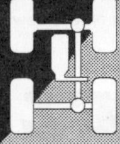

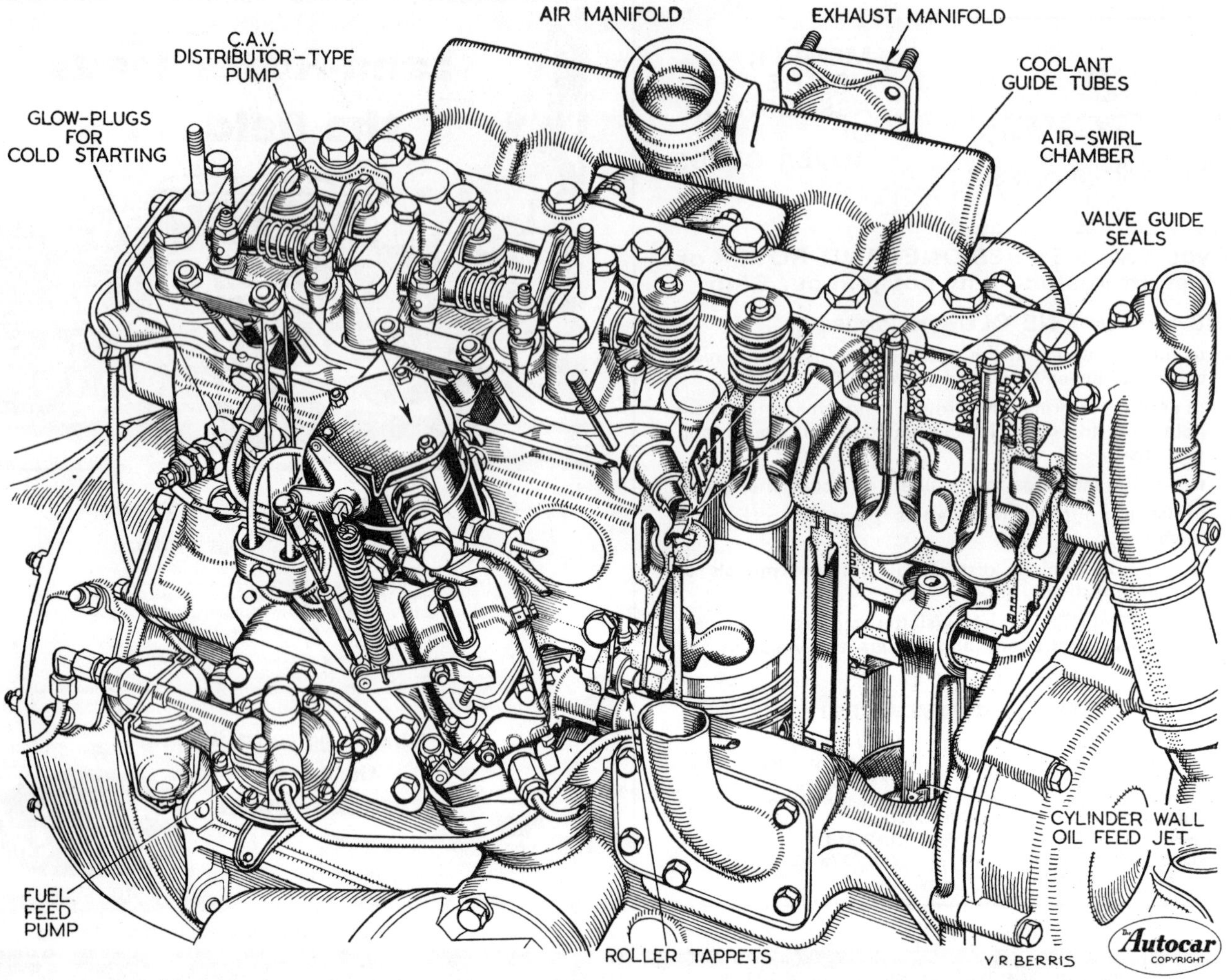

AIR MANIFOLD
EXHAUST MANIFOLD
C.A.V. DISTRIBUTOR-TYPE PUMP
COOLANT GUIDE TUBES
GLOW-PLUGS FOR COLD STARTING
AIR-SWIRL CHAMBER
VALVE GUIDE SEALS
CYLINDER WALL OIL FEED JET
FUEL FEED PUMP
ROLLER TAPPETS
V.R. BERRIS
The Autocar COPYRIGHT

A three-bearing crankshaft, wet liners, Ricardo Comet V combustion chambers and C.A.V. distributor-type injection pump are features of the new engine

LAND-ROVER GOES DIESEL

Alternative Two-litre Engine for Greater Economy

TO widen the appeal of the ubiquitous Land-Rover, the Rover engineers have produced a very efficient alternative diesel engine. Its speed range is very close to that of the petrol engine, developing 52 b.h.p. at 3,500 r.p.m. (petrol engine output is 52 b.h.p. at 4,000 r.p.m.), and the engines are interchangeable on the 7ft 4in and 9ft 1in wheelbase models. It will be recalled that the wheelbase of each of these models was increased by 2in at the last London Motor Show. For the moment, the diesel engine cannot be fitted as a replacement for the earlier shorter wheelbase models.

The diesel engine is completely new and is not a conversion from the petrol engine, the only common feature being the centre distance of the timing chains between crankshaft and camshaft.

The combustion system utilizes a swirl chamber of the Ricardo Comet V type, comprising a modified spherical cell in the head, with a tangential hole connecting to two spectacle-pattern recesses in the piston. Although slightly improved fuel consumption can be achieved with direct injection, i.e., with toroidal-shaped corabustion chambers formed in the piston, such a design is not easy to accommodate in small cylinder sizes, and on overall balance the swirl chamber offers greater benefits.

A disadvantage of the swirl chamber is that a substantial proportion of the air at high velocity must pass through the restricted passage, and the combustion products must also pass out through the same passage at still higher velocity. Against this, a greater proportion of the air retained in the cylinder can be used, and thus permit a higher mean effective pressure at the clean exhaust limit. Also, since the injector is at one side of the head, larger valves, with freer entry, permit higher volumetric efficiency, with the ability to hold the m.e.p. at higher engine speeds.

The crankcase is an iron casting with an integral water gallery, in which there are four jets to direct the coolant round the cylinders. Wet liners are used, with a top retaining flange, and square-section neoprene sealing rings housed in the cylinder block at the lower end. The three-bearing crankshaft is a substantial heat-treated forging with $2\frac{1}{2}$in diameter by $1\frac{1}{16}$in wide bearings at the front and centre; the rear is much wider—$1\frac{3}{8}$in.

55
50
45
40
35
30
25
20
15
B.H.P.
110
100
90
80
70
TORQUE LB.FT. B.M.E.P. LB.SQ.IN
·5
·4
SPEC. FUEL CONS. LB./B.H.P. HR.
B.M.E.P.
TORQUE
B.H.P
FUEL
1,000
2,000
3,000
R.P.M.

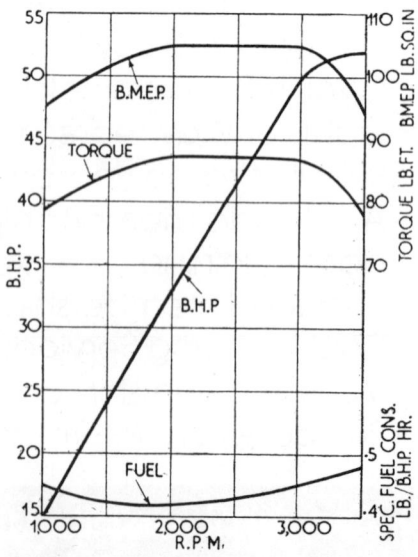

Nett power curves of the new engine at clear exhaust limit

Land-Rover Goes Diesel . . .

The big ends are 2½in diameter by 1⅟₁₆in wide; the conrods, split horizontally, can be withdrawn through the cylinder bores. Copper-lead thin wall bearings with tin overlay are used.

The cylinder head is an iron casting, without seat inserts; it has controlled water passages to ensure a free flow of coolant round the injector pockets and valves.

The case-hardened steel camshaft incorporates a spiral gear for driving the injection pump. Inclined roller-type tappets, of a novel design, are claimed to provide high accelerations with minimum of wear on the cams. Each roller runs in a tin-plated bronze shoe which in turn slides in a steel tappet guide. The camshaft is placed relatively high in the cylinder block, and connects to the rockers with short, solid push-rods.

The Duplex roller chain is tensioned on the slack side by a hydraulic tensioner fed from the engine lubrication system; a rubber damper pad on the taut side eliminates chain thrash.

There are three compression rings (the top one parallel-faced chrome plated, the

The new diesel is completely interchangeable with the petrol engine

Nearside view of the engine showing manifolds, belt-driven auxiliaries and crankshaft damper

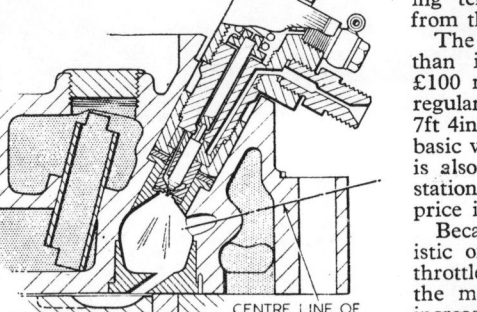

The form of the combustion swirl chamber and spray characteristics of Pintaux injectors

CENTRE LINE OF GLOW-PLUG

two lower ones taper faced), and one oil control ring above the gudgeon pin. Additionally, there is a groove in the skirt below the pin in which a second oil control ring can be fitted later in the life of the engine in the event of an increase in oil consumption. The connecting rods incorporate an oil jet from the big end bearings to lubricate the thrust side of the cylinder wall.

The injection pump is the new C.A.V. DPA distributor type. It is developed by C.A.V. from the American Roosa-Master, and is a single-cylinder, opposed plunger design with a rotating shank which embodies a distributor to each cylinder in turn. For small engines this type of pump has many advantages over the earlier, in-line jerk type, being compact and containing no ball or roller bearings, gears or highly stressed springs. It has an all-speed mechanical governor which operates during normal running and when the power take-off auxiliaries are in use.

Fuel is fed to it from an AC-Delco type diaphragm feed pump, also operated from the camshaft by an eccentric lobe, and there is a C.A.V. paper element filter in the system.

Pintaux-type nozzles are used with the Ricardo Comet V swirl chamber. This design incorporates two feed holes, one coaxial with the injector and one drilled at a tangent. When the needle valve first starts to lift, the side hole is opened, and fuel is sprayed into the hottest zone of the compressed air when starting from cold, the main jet still being partially closed by the pintle.

At normal running speeds, the needle lifts fully and only a small proportion of fuel is delivered through the tangential side hole into the direction of air swirl. These features provide for complete combustion over a wide range of fuel/air mixtures, and ensure most economical operation.

Glow-plugs, fitted as an additional aid to starting in extra low temperatures, would not be required in normal operating temperatures. They are controlled from the instrument panel.

The diesel version weighs 108 lb more than its petrol counterpart, and costs £100 more. It can be supplied for the regular Land-Rover; the prices are— 7ft 4in wheelbase £715; 9ft 1in wheelbase basic vehicle £790, and de luxe £810. It is also available for the short wheelbase station wagon—basic price £785, total price in U.K., £1,178 17s.

Because of the diesel engine's characteristic of better fuel consumption at part throttle than that of the petrol engine, the manufacturers claim a 50 per cent increase in average m.p.g. The engines are immediately available in the U.K. and will be available later in the year overseas.

SPECIFICATION

No. of cylinders	...	4 in line	Compression ratio	...	19.5 to 1
Bore and stroke	...	85.72 × 88.9 mm (3⅜ × 3½in)	Injection pump	...	C.A.V. DPA distributor type
Displacement	...	2,052 c.c. (125.2 cu in)	Injection nozzles	...	C.A.V. Pintaux type
Valve position	...	In-line O.H.V., push rods	Fuel filter	...	C.A.V. paper type
Max b.h.p.	...	52 at 3,500 r.p.m.	Oil filter	...	AC-Delco full flow
Max b.m.e.p.	...	105 lb sq in at 2,000 r.p.m.	Fuel feed pump	...	AC-Delco diaphragm type
Max torque	...	87 lb ft at 2,000 r.p.m.	Cooling system	...	Pump, fan and thermostat

PROVIDING a clear 50 per cent. improvement in fuel consumption, the new Rover 2-litre oil engine, which was introduced last June, and is available as optional equipment in all versions of the Land-Rover 4 x 4 chassis (except the 107-in. wheelbase station wagon), gives almost the same road performance as the 2-litre petrol unit fitted as standard. During a 1,000-mile test, engine power was proved repeatedly, both on the road and across rough country and farmland, and the overall fuel-consumption rate was 29.5 m.p.g., whilst the engine-oil consumption rate was more than 4,000 m.p.g.

Criticism has been levelled at the engine in respect of its noise, but on the test bed it is no noisier than other units of this size. As installed, however, the new unit suffers the disadvantage of being shrouded by flat metal panels which act as a sounding box. In addition, hard mountings tend to accentuate the vibration and noise, particularly at idling speed.

So far as the model which formed the subject of our test is concerned, one of the more noticeable improvements incorporated recently has been the adoption of recirculating-ball steering, which makes the steering much lighter than it was previously. Other modifications include the use of wider and more robust road springs; more substantial exhaust-system mountings; and better body draught-proofing.

The vehicle supplied was a short-wheelbase Regular model and optional equipment fitted included heater and demister, flashing direction indicators, 6.50-16-in. cross-country tyres, rear seats and link-type floor mats.

The second visit to the A... heavy rain had fallen, wit... traction was maintain...

By John F. Moon, A.M.I.R.T.E., *The Commercial Motor,* **and**
T. Hammond Cradock, *Farm Mechanization*

SAME UBIQUITY—
Greater Economy

New Rover 2-litre Oil Engine Shows Clear 50 per cent. Improvement in Fuel Consumption Without Reduction in Power

(*Left*) *Unloading sacks o... objects of combine harves... possible, for which the La... ground and its rears just ... at Bagshot Heath. (Abo...*

(*Left*) *Collecting bales of oat straw with two trailers hitched in tandem. The Land-Rover was driven at its lowest speed during this field operation.* (*Right*) *Breasting the top of a 1 in 10 rise, the 4 × 4 pulls the 1½-ton engine-driven Allis-Chalmers combine harvester.*

e Course at Bagshot Heath was made after
seen above. Despite the mud and slime, full
Land-Rover remained fully controllable.

A non-stop "bottom-low" climb of the 1 in 1.73 (30°) slope at the F.V.R.D.E.
proving ground was made despite the slippery condition of the timber baulks
which surface this gradient. The vehicle started from rest on 1 in 2.

an Allis-Chalmers All-Crop combine harvester. One of the main
t the grain from the field to the farmyard storage as quickly as
entirely suitable. (Above, left) With its front wheels clear of the
, the Land-Rover shows its high-speed cross-country performance
he pavé track at the F.V.R.D.E. often caused the rear springs
, but no serious damage resulted.

With a half-full fuel tank the vehicle weighed 1 ton
7¼ cwt. and a test load totalling 7 cwt. was added.
This is nearly 2 cwt. less than the permissible payload
with driver and two passengers aboard, but is the
standard rating for cross-country operation, of which
much of the test was to consist.

Before leaving the Rover works, the fuel tank was
topped up to overflowing and thereafter a careful check
was kept on all fuel added. The journey to London
was completed at an average of almost 40 m.p.h.,
cruising mostly at about 50 m.p.h. At this speed the
engine noise and the whine of the transmission were
almost drowned by the tyre roar created by the cross-
country lugs: indeed, at any speed above 30 m.p.h. tyre
noise tended to obscure mechanical sounds.

Off the Road

FOR the first test of the series the Land-Rover was
taken to the Ministry of Supply's Fighting Vehicles
Research and Development Establishment. Tackling
the test slopes first, the 1 in 4 and 1 in 3 gradients were
climbed non-stop in bottom gear, high ratio, and on
both slopes satisfactory restarts were made from a

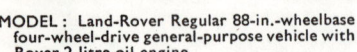

MODEL : Land-Rover Regular 88-in.-wheelbase four-wheel-drive general-purpose vehicle with Rover 2-litre oil engine.

WEIGHTS :

	Tons	cwt.	qr.
Unladen (kerb weight)	1	7	1
Payload		7	0
Driver, observer, etc. ..		4	0
	1	18	1

DISTRIBUTION :

Front axle		17	1
Rear axle	1	1	0

ENGINE : Rover four-cylindered indirect-injection oil engine ; bore 85.7 mm. (3.375 in.) ; stroke 88.9 mm. (3.5 in.) ; piston-swept volume 2.052 litres (126 cu. in.) ; maximum net output 52 b.h.p. at 3,500 r.p.m.: R.A.C. rating 18.2 h.p.; maximum net torque 87 lb.-ft. at 2,000-3,000 r.p.m.

TRANSMISSION : Through 9-in.-diameter single-dry-plate clutch to four-speed synchromesh gearbox and two-speed transfer box, thence by single propeller shafts to the fully floating spiral-bevel front axle and the semi-floating spiral-bevel rear axle.

GEAR RATIOS : Main gearbox, 2.996, 2.043, 1.377 and 1 to 1 forward ; reverse 2.547 to 1 ; transfer box, 2.888 and 1.148 to 1 ; axle ratio 4.7 to 1.

BRAKE : Girling hydraulic system with leading-and-trailing-shoe units at all wheels. Hand brake linked mechanically to internal-expand-

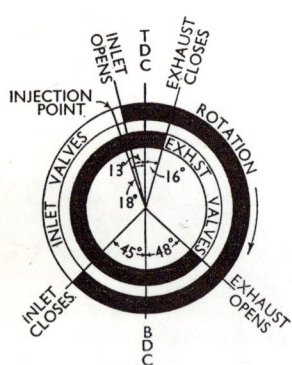

COMPRESSION RATIO 19·5:1
FIRING ORDER 1·3·4·2
VALVE CLEARANCES 0·010"

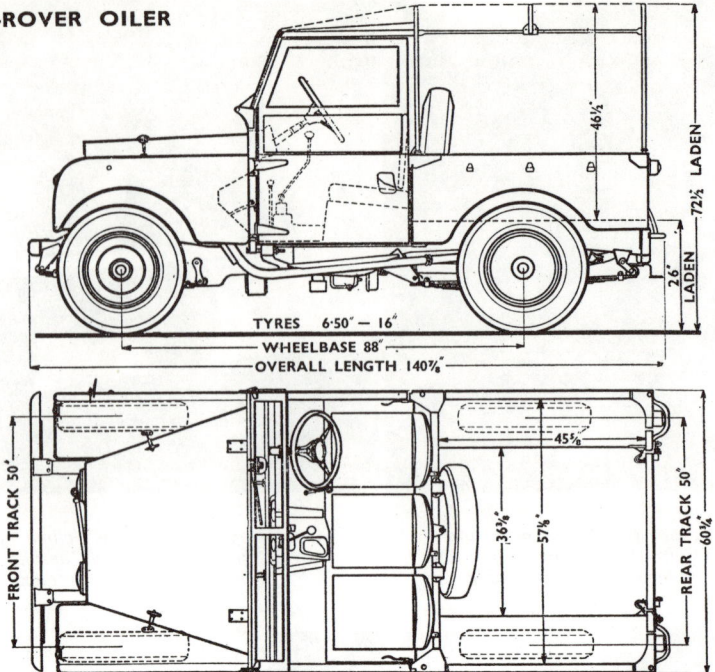

ing transmission brake on rear of transfer box. Diameter of drums, 10 in. ; width of facings, front, 1.5 in., rear, 1.5 in. ; total frictional area 104.7 sq. in. : that is 54.5 sq. in. per ton gross weight as tested.

FRAME : Welded box section with three box-section cross-members welded in position.

STEERING : Burman recirculating-ball worm and nut.

SUSPENSION : Semi-elliptic springs with telescopic dampers at both axles.

ELECTRICAL : 12v. current voltage - control system with 120-amp.-hr. battery.

FUEL CONSUMPTION : (a) laden, low speed, 37.7 m.p.g. at 29.25 m.p.h. average speed ; (b) laden, high speed, 31.2 m.p.g. at 40 m.p.h. average speed ; (c) unladen, low speed, 40 m.p.g. at 29.6 m.p.h. average speed; (d) un-laden, high speed, 33.8 m.p.g. at 39.5 m.p.h. average speed ; that is 72.1 gross ton-m.p.g.

as tested (a) and 59.7 (b), giving time-load-mileage factors of 2,108 and 2,388 respectively.

TANK CAPACITY: 10 gal., laden range approximately 350 miles.

ACCELERATION : Through gears, 0-20 m.p.h. 5.75 sec.; 0-30 m.p.h., 11.5 sec.; 0-40 m.p.h., 21.25 sec.; 0-50 m.p.h., 37.75 sec.; top gear, 10-20 m.p.h., 9.75 sec.; 10-30 m.p.h., 18.75 sec.; 10-40 m.p.h., 30.25 sec.; 10-50 m.p.h., 47 sec.

BRAKING : From 20 m.p.h., 19.5 ft. (22.2 ft. per sec. per sec.); from 30 m.p.h., 46.5 ft. (20.9 ft. per sec. per sec.)

WEIGHT RATIO : 1.35 b.h.p. per cwt. gross weight as tested.

FORWARD VISIBILITY : To within 12 ft. of front bumper at ground level on centre line.

TURNING CIRCLES : 42 ft. both locks.

MAKERS : The Rover Co., Ltd., Solihull, Warwickshire.

---

standstill in second gear, low ratio. Facing down these hills, smooth reverse re-starts were made also and engine braking was such as to allow both slopes to be descended without use of the brakes at 5 m.p.h.

Second low was required for a non-stop climb of the 1 in 2 slope and a smooth bottom-low re-start was made on it. Similarly, on the way down—for which the brakes were lightly applied—the vehicle was stopped and a successful reverse re-start made.

This slope, like the 1 in 1.73 gradient, is faced with timber slats for tracked vehicles, and although they present no problems when dry, they make traction difficult when wet. Nevertheless, a confident climb of the 1 in 1.73 slope was made in bottom low, under both wet and dry conditions, to the surprise of members of the Army.

As later tests were to prove, rough-surfaced slopes—wet or dry—of up to 1 in 2 can be tackled with ease.

To try the suspension the 4×4 was taken over the three suspension courses, each of which is 300 yd. long. The tracks are paved with pavé, and with 1-1½-in. and 2-2½-in. raised setts staggered so as to vary the periodicity of the shocks transmitted to the suspension.

The vehicle handled well over these surfaces, considering the simple suspension, although over the 2-2½-in. setts the highest speed that could be maintained was 15 m.p.h. The vehicle was then only just controllable, being turned at times through 45° to the direction of travel. As a result of four runs over each of these tracks the total damage sustained was a loose throttle pedal.

On the Alpine Course at Bagshot Heath, the Land-Rover was put through its paces under both wet and dry conditions on separate occasions. Despite the severity of the circuit, most of the trial was carried out in high auxiliary ratio without front-wheel drive being engaged.

It became virtually impossible to find any part of the ground that could not be reached with the Land-Rover and even though, when crossing very wide ditches, the vehicle sometimes became temporarily "grounded," either the front or rear wheels would maintain sufficient traction to enable the 4×4 to be shunted out easily.

Often all four wheels left the ground completely. It was during one jump that the front axle was bent upwards to the left of the differential pot. The passenger was also bent! An improved grab handle for the front-seat passengers is needed, there being little to grasp during high-speed runs over rough ground.

On the Farm

ALTHOUGH the Land-Rover has a power-take-off shaft and a belt pulley (optional extras), which can be used to drive a wide range of stationary and mobile machines, including mowers, sprayers and saw benches, its main value as a farm vehicle is its ability to haul and carry on roads and across rough country.

For certain operations it can act as a tractor by pulling what are known as trailed implements (those usually hitched by an orthodox drawbar pin), but there are many operations for which it is unsuitable because the available implement has been designed for three-point mounting and hydraulic control, and to be used almost in unit construction with the tractor. There are also implements designed to be used under the belly of the tractor or mounted on the front axle, and these cannot be used with the Land-Rover.

This means that the vehicle cannot be a complete substitute for a tractor on the majority of British farms, but it is suitable as a complementary unit to practically any system of farm mechanization. It can carry a 10-cwt. load under most conditions and there is sufficient tractive effort available for it to assist in pulling machines out of trouble.

Our farm tests were carried out on Chested Farm, Chiddingstone, Kent, where Mr. A. C. Baynes and his son farm 125 acres, including about 50 acres of grain and 25 cows in milk. Mr. Baynes, who was known to the world a few years ago as the one and only Stainless Stephen, is operating his second petrol-engined Land-Rover, so that he and his son Ian were able to make valuable comparisons.

The first remark made by Ian Baynes when he drove the test model was: "The steering is lighter than mine." Mr. Baynes, Snr., took over in the field while Ian and we prepared to do some hard manual labour, he expressed approval of the hand throttle.

The manual labour consisted of loading bales of oat straw on to two trailers hitched in tandem to the Land-Rover. These bales weighed about 50 lb. each and lay scattered about the field where dropped by the pick-up baler. They are not the sort of thing to run about with and we soon discovered that the low-bottom gear was just a shade too fast to allow the vehicle to keep moving all the time. Had a "creep" speed been available, we might have persuaded the driver to lend a hand at loading while the Land-Rover virtually drove itself.

Full marks were awarded for the way in which the vehicle handled the loaded trailers out of the field and on the road. There were ample power and tractive effort, although the two-wheeled trailers tended to lift the rear of the vehicle and thus reduce the weight (and adhesion) on the rear wheels.

The next test was to pull an Allis-Chalmers All-Crop combine harvester which cuts, threshes and bags the grain in one operation. Here, again, the low-bottom ratio proved to be a shade too high for the best results, because the engine had to be throttled down almost to a tick-over to keep the forward speed of the combine low enough for it to deal with the crop.

But even at almost idling speed the engine pulled the combine up a 1 in 10 slope. The combine weighs 3,360 lb. and given a low ratio which would allow the engine to be run at half throttle or more, the Land-Rover would be able to haul it under any condition suitable for the farm tractor.

Carting sacks of grain from the combine, the Land-Rover came into its own again and a load of over 2 tons was handled with ease across fields and on hills.

On the Road

DURING the road section of the tests, most satisfactory acceleration times were recorded. The braking system was powerful, the stopping distances being marred by skidding because of the reduced surface contact of the tyres.

Two hand-brake tests were made: one with rear-wheel drive only in operation, during which an average Tapley meter reading of 41 per cent. was obtained, and one with four-wheel drive engaged, when 53.5 per cent. was recorded.

Tests for coolant-temperature rise and brake fade were conducted on a ¾-mile hill with an average gradient of 1 in 10½. The coolant tests showed the system to be adequate for prolonged hill working in ambient temperatures of less than 100° F. and only slight brake fade was recorded after a descent in neutral.

Throughout the tests the Land-Rover handled extremely well (despite the bent axle) and it was found to have a maximum speed of 53 m.p.h. The seating is extremely comfortable, particularly on long journeys, and ventilation is adequate. In extremely hot weather further ventilation can be obtained by removing the cab doors completely.

Minor criticisms concern the absence of a near-side rear-view mirror and the poor direction indicators. Front and rear winkers are combined with the side and tail lamps, and the small green repeater light on the instrument panel is almost impossible to see during daylight.

The hood is waterproof, but is a little complicated to remove and refit. The rear flap also is somewhat difficult to raise or lower, three minutes and two minutes, respectively, being required, for these jobs. On the open road at speed, with the flap raised, exhaust gases are drawn into the body: this can be avoided if the front ventilators are opened.

On the few fairly cold mornings that occurred during the time of the test, engine starting presented no problem. The best technique is to use the heater plugs for a few seconds before actuating the starter motor. The combined starter and heater switch makes this procedure particularly simple.

In the Workshop

AS the first of a series of maintenance tasks, I raised the bonnet (8 seconds) on its hinged support. The bonnet can, even if the spare wheel is mounted on it, be raised easily. Removal of the wheel, however, takes only 30 seconds and replacement 1½ minutes.

With the bonnet up it took 6 seconds to check the water level, 18 seconds to verify the engine-oil level, 33 seconds to test the battery electrolyte levels and 2¾ minutes to check the air-cleaner oil level.

A dipstick in the top of the gearbox is reached through a hinged plate in the top of the transmission tunnel. The oil level was checked in 35 seconds. The oil-filler orifice is large and easy to reach.

The transfer-box oil level, however, is not so easy to check, there being a very small level plug in the back of the unit which makes it impossible to see or feel the level. Other than by topping until oil starts to flow out through the hole, the best way of checking the level is to use a matchstick, and in this way the task was carried out in 45 seconds.

The front-axle oil level is simple to check, there being a large filler and level plug in the front of the casing which can be removed and replaced in 40 seconds. The rear-axle plug is less accessible and this check took 1¾ minutes.

Returning to the front axle, the oil levels in the swivel-pin housings were each verified in 45 seconds, these oil baths being distinct from the main axle lubrication system.

The brake-fluid reservoir is adjacent to the fuel tank, beneath the driving-seat cushion, and the fluid level was checked in 30 seconds. This time included the removal and replacement of the cushion. Unfortunately, there is no strap to hold up the seat-box lid, which has to be held in one hand. This applies also when filling the fuel tank.

At the Rover works several engine tasks were conducted and to make these easy the bonnet was removed completely (30 seconds). All four injectors were then taken out in 12 minutes (this time includes detaching the air cleaner), it being easier to remove all four injectors at once, as this avoids the necessity of taking off the return pipe.

Having cleaned the nozzles and fitted new seating washers—this is most important—the injectors were replaced in 17½ minutes, but before doing this the lift-pump sediment bowl was removed and replaced in a minute.

To remove the main fuel-filter element took six minutes, because it was necessary to withdraw the right-hand battery to pull the filter bowl clear of its location. Reassembly of the filter also occupied six minutes, including battery replacement time.

The complete fuel system was bled at the injection pump in 1 minute 20 seconds, and the bonnet replaced in 45 seconds.

As a final check, the times required to remove and replace the spare wheel from its alternative stowage point behind the front seats were found to be 23 and 40 seconds, respectively. The wheel was not carried here during the tests because of the intrusion upon payload space.

The 88-in.-wheelbase Land-Rover with oil engine sells for £730, this being £100 more than the petrol-engined model. No purchase tax is applicable unless the vehicle is purchased as a station wagon, in which case the basic price is £705 for the petrol model, plus £353 17s. purchase tax.

Land-Rover

driving around

with walt woron

IF YOU CLASS YOURSELF as one of the carriage trade, you're not apt to be too interested in anything as utilitarian as a Land-Rover. On the other hand, if you ordinarily dress in blue jeans, coveralls, or wear a straw hat, you might find a specific use for this machine.

A week of driving the Land-Rover 109 into the rugged Sierra mountains convinced me that the sales department of the Rover Co. Ltd. of Solihull, Warwickshire, England is right in billing the truck as ". . . essentially a vehicle of action, having been designed and built to tackle a very wide range of duties." In one of its four versions on three different wheelbases (88-inch pickup and wagon, 107-inch wagon, and 109-inch pickup) it is being used by a practically endless list of customers: farmers, mountaineers, explorers, prospectors, game wardens, policemen, hunters, fishermen. Because it can be rigged with a power take-off, a winch, and a hitch it can be used to fell trees, spray crops, cultivate, or tow other vehicles. It will do practically anything you ask of it—and then some.

The Land-Rover 109 looks bigger than it is—probably because it stands almost 20 inches above the average '57 American car. Yet its overall length, including a 73-inch bed, is only 10 inches more than the length of a Nash Metropolitan.

Much galvanized steel has been employed throughout—virtually wherever there might be hard usage. It's used on the door edges, as trim on the pickup bed, as rubstrips on the benches and bed floor. There's nothing fancy about it, nothing to delude you into thinking you're getting something you're not, though finer appointments are optional. Screwheads show, hinges are out in the open, welds indicate how ruggedly it was assembled. There is another reason, too. It facilitates removal and/or servicing. Remove a few bolts, the top comes off. Loosen a coupe of toggle bolts, the windshield can be removed. Open the doors wide and lift them completely off their hinges. The body can slowly and fairly easily be stripped down to the bare chassis.

Further indication of the inbuilt ruggedness is in the Land-Rover's chassis; two heavy, boxed-in frame sections are used with four crossmembers. The front is suspended independently by heavy semi-elliptics on a solid axle, with the wheels hanging from enclosed ball-and-socket joints. The rear is suspended with semi-floating shafts and semi-elliptics. With such rugged springs and the use of non-adjustable tubular shocks all around, a soft ride is hardly to be expected. What you get is a tolerable ride for short distances but much choppiness under any condition. On the other hand, you can take corners with little fear of tossing your passengers about.

To get into the Land-Rover, you twist and pull up the handle, open the flat 32- by 43-inch doors, and step up a long two feet from the ground. It's a stretch for a man, but a woman either needs a strong arm to lift her up or the car has to be parked next to a curb. A swing-out platform step would be a practical addition to the long optional list. Once in, driver and passenger are fairly comfortable on the removable padded seats. A third passenger can be accommodated on the middle seat, but legroom is restricted by the gearshift controls. Otherwise, there's lots of legroom, headroom, hiproom and footroom.

You won't be bothered by having to read many instruments, for the only standard ones are the speedometer, ammeter and fuel gauge. (On the steep roads into Sequoia National Park I wished I'd had the temperature gauge, for I didn't know if the engine was overheating without an occasional stop and look-see.) All instruments are in a central panel, where they are not at all easy to read in a quick glance.

Sitting up as high as you do, you get a good view of the road ahead and the terrain around you. It would help, though, if the

74

rear sliding window were made wider, for your view to the left rear is quite restricted unless you depend solely on your side view mirror.

When you consider the fact that the Land-Rover is pulled along by a four-cylinder engine rated at only 52 horses at 4000 rpm, you don't wonder that a comfortable highway cruising speed is only 55-60 mph. On a slight incline your speed will start to drop and you'll have to shift down to third, and be content with a maximum speed of 50 mph.

Despite the low horsepower rating of the 122-cubic-inch, F-head engine, you have a healthy torque output of 101 pounds-feet at 1500 rpm. By the judicious selection of gears, this torque can be multiplied so that the lowest overall ratio is 5.40 in fourth gear and a fantastic 40.69 in first gear. This is accomplished through the main gearbox, the transfer box and 4.7 to 1 rear axle.

Paradoxically, the most complicated and most interesting feature is one and the same: the gearbox control. The main gearbox is shifted with the floor-mounted control, requiring double-clutching from first to second since these two gears are not synchromesh. A yellow knob located beside the main gearshift controls two-wheel and four-wheel drive: up for two-wheel drive on the open highway; down for engaging the front-wheel drive so that added traction can be gained on soft surfaces.

The transfer box gives two ratios in the output from the main gearbox. Normally, it is kept in the high position, but for traversing muddy or sandy surfaces, pulling a heavy load, or climbing a steep grade, low transfer is used. Shifting from high to low is done with the clutch, although the car must be stationary, while upshifting can be done at any time. When low transfer is engaged, four-wheel drive is automatically engaged also; it disengages when shifting back to high transfer.

Servicing or working on the various components of the engine presents little difficulty. The raised hood reveals an easily removable intake valve cover, out-in-the-open 14 mm plugs and the Lucas 12-volt distributor, a Solex downdraft carburetor getting its fuel from an S.U. electric fuel pump mounted on the firewall, the oil filler neck at the left front, the oil dipstick below the intake manifold, and an easy-to-get-at exhaust valve cover on the side of the block. Adjustment of the valves could be done even while the engine is hot with no danger of burning yourself on the exhaust manifold.

Don't expect to win any drag races—even against a sick Greyhound—but on the other hand, don't let the steepness of any hill or the softness of the ground underfoot faze you. As Henry Henkel of Rootes Motors in Los Angeles told me before I took off, "When you think it won't make it, that's when it starts to go." He should know. He took one on a 2300-mile round trip to LaPaz in Lower California, which includes some of the roughest terrain you will encounter west of the Rockies.

The less often you have to take on fuel, the more you'll like it; for two reasons. One naturally, involves your billfold. The other is that the 12-gallon tank is located under the far right seat. You have to lift off the seat pad, raise the locker lid, and remove the cap before the attendant can start pumping in the regular fuel required by the 6.7 or 6.9 compression ratio engine. If he's sloppy and spills some gasoline, or if the cap is not secured tightly, you're going to get annoying gas odors in the cab. If you are not carrying a load on the 1200-pound capacity bed, you can expect better than 15.4 mpg around town, 12.8 in the mountains and over 20.4 mpg on the open highway. These were the averages I got on a brand-new Land-Rover, so are under what you should anticipate when the truck is properly broken in.

Should you buy one? Not as a replacement for a car, nor as a substitute for a two-wheel drive pickup. But, if you need a four-wheel drive vehicle in your line of business or are that extreme a hobbyist, it could be.a wise choice. At a $2974 Port of Entry price, the 109 costs more than a Willys Four-Wheel Drive Jeep, a Forward-Cab Willys, and a Dodge Four-Wheel Drive pickup, but less than a Chevy Power Wagon. The smaller version (88-inch wheelbase pickup) sells for $2561. I'll leave the choice up to you.

THE INCH WAR

It may have looked like a normal Land Rover, but the Rolls-Royce powered military prototype had a variety of changes under the skin . . . including an inch longer chassis. Tony Hutchings reports

It is well known that Rover looked closely at the American wartime Jeep when designing an agricultural workhorse as a stop gap in 1947. The original centre-steering prototype was built on an army surplus Jeep chassis, which was later fitted with Rover axles and transmission, the first items made specifically for the Land-Rover.

As the Jeep had an 80in wheelbase, it followed that the Land-Rover 'inherited' the same dimension. The single prototype was soon followed by a batch of 48 pre-production galvanised chassised vehicles which, although very different from the prototype, still retained the 80in wheelbase. And that wheelbase remained unaltered when quantity production began in July 1948.

During 1950, however, military trials were held in the Chobham area with numerous manufacturers entering competitive 4 × 4 vehicles against the B40 Rolls-Royce-engined Champ. Rover entered two vehicles, the standard 80in 1595cc Land-Rover and a special Rolls-engined 81in wheelbase model. The latter was requested by the Army, who were very interested in using the B40 as a standard engine in various different types of vehicles. When the Army approached Rover to carry out this modification, Rover arranged with Hudson Motors Ltd, to install and develop the 2.8-litre B40 Mk 2B engined vehicle. Standard vehicles were supplied off the production line, complete with engines which were later returned to the works . . . at the time the Rolls engines cost more than a complete Land-Rover!

The 81in specification differed considerably from the standard machine. Power and torque were increased by 50 and 75 per cent respectively, but the transfer gear ratios were changed in order to keep the tractive effort at approximately the same figure as the standard vehicle, thus giving the 81in a much higher maximum speed. Other differences between the two included modified frame cross members; a new bell-housing, which was made to accommodate a larger, 10in clutch; repositioned clutch and brake linkages; a lengthened front prop shaft; relocation of the battery under the passenger seat instead of under bonnet, while the size of the radiator was increased and the cooling system pressurised to 10lbs psi against 5lbs psi.

Because of the size of the B40 engine it was necessary to raise the bonnet using rubber buffers; the bonnet top was 'dressed' to fit the radiator header tank and a circular hole had to be cut in the forward part of the bonnet to accommodate the radiator cap – a distinctive mark of this particular model. The front bumper was raised 1½ inches to allow the starting handle to line through with the dog on the end of the crankshaft. The unladen weight was 25cwt as against 23cwt of the standard vehicle. Rolls-Royce records show that 54 B40 2B engines were produced to specification PL6059 and numbered 534, 556-605, 609 and 610.

Because of the tightness of the under bonnet fit it was not possible to fit a crankshaft damper – thus this model suffers from some vibration. Presently only two 81in vehicles are known with the Rolls engine insitu, chassis numbers R061 04546 (owned by Tim Ralphs) and R061 04618. The latter Land-Rover, registered TAB 767, was the last of the batch of 34

Top: The 81in at rest. Above: Rolls-Royce engine insitu, giving 50 per cent more power than the standard offering

built in November 1949, with engine no 596, and was sold by the Ministry of Defence in 1953. It was bought by Messrs Belliss and Morcom.

In 1957 it was bought by a Birmingham garage owner (a Rolls-Royce car owner) and was used very little. When it was next offered for sale, in 1977, Ian Sparks of Birmingham did not know exactly what it was on first sight, but was lucky enough to learn about the model from David Moss of the Land-Rover Register during the three days of purchasing negotiation. Subsequent inquiries revealed TAB's history and a 'factory' manual and other technical literature and information were unearthed.

TAB 767 was painstakingly restored by Ian during 1977-8. It is a very pleasant vehicle to drive and the acceleration is really outstanding. The lack of a crankshaft damper means body vibrations occur at about 40mph but this speed is soon passed and Ian confirms the top speed is 80mph with a comfortable cruising speed of 50mph. A most impressive vehicle that handles just like a Series III Land-Rover, it has won Ian many Concours D'Elegance events over the past three years and causes interest wherever it is seen, particularly should the bonnet be raised.

Oh, and by the way, the vehicle that came out best at the 1950 Military trials? The standard 1595cc 80in, of course!

Buying A Land-Rover

For a classic with a difference why not buy the world's number one multi-purpose vehicle? By John Williams.

Why buy a Land-Rover? There are those who own these vehicles for fun, and there are plenty who own them for their working capabilities, *and* for pleasure. For a lot of owners the particular appeal of the Land-Rover is its rugged simplicity, and others will add that you don't worry about a Land-Rover if it gets the odd scratch on the paintwork, or if it gets dirty – which is more than can be said for most classic cars – as it was made to be used, and used almost anywhere.

Are Land-Rovers classics? *Practical Classics* has never presumed to define what is or is not a classic car or classic commercial vehicle, but Land-Rovers – or at least the early ones – can claim a number of useful qualifications. They were the leaders in the field of four-wheel-drive multi-purpose vehicles for many years (it is arguable that they still are the leaders). They were first produced in 1948 and were already enjoying the support of a number of owners clubs by the mid-1950s.

Their success, and their significance, can hardly be measured by the fact that only 30,000 of them had been made by 1961 (and the majority exported) but there is hardly a country in the world which is not familiar with the Land-Rover by now.

Although designed to cope with light agricultural tractor work the Land-Rover's reputation was built upon its hard working 'go-anywhere' capabilities.

A little history

After the Second World War steel was in short supply and motor manufacturers were being severely rationed. The Rover Company Ltd, which had been making cars since 1904 and which had gained a reputation for quality, soon realised that some of its plans for post-war models would have to be postponed. The company could not expect to survive on the limited production of cars which its steel allocation would allow, and there was little hope of in an increased allocation as this was related to export achievements; luxury cars were not a high priority requirement of the export markets during the post-war years.

Series I Land-Rovers were built in a variety of wheelbase sizes: 80", 86", 88", 107" and 109", but the Series 11 models were 88" and 109" only. This picture shows 107" and 80" models; the shortage of load space in the latter is obvious...

Maurice Wilks, who was the head of design at Rover at the time, has been credited with the original idea for the Land-Rover. The story goes that Mr Wilks had a 250 acre estate in Anglesey on which he needed a vehicle which would not only keep going over a variety of ground conditions but would tow, plough, do various other agricultural tasks and drive other machinery. First he used an ex-WD half-track Ford truck which he found somewhat unwieldy. Another ex-WD vehicle, the famous wartime Willys Jeep, was the next to be tried and this might have been more acceptable had it not been an imported product. Maurice Wilks came to the conclusion that there was probably a world-wide market for a versatile, go anywhere, Jeep-like vehicle and at the same time his brother Spencer Wilks (Rover managing director) was looking for a stop-gap project to utilise spare factory capacity until such at time as the planned post-war model programme could be put into effect.

Development went ahead quickly and it was not long before Land-Rover production outnumbered the Rover cars being made.

Features and modifications

It is thought that a small number of prototype Land-Rovers were built followed by 48 pre-production models for evaluation purposes. These 48 were numbered L or R 01 to 48

Mechanical components are capable of lasting very well in normal road use but wear is to be expected if there is much use on this type of terrain.

Access to the engine is not outstanding and with that spare wheel on the bonnet you will want to be sure that the bonnet support is in good shape.

The front hubs can incorporate a device which enables the wheels to be engaged or disengaged from the transmission so as to minimise wear in the forward transmission components when four wheel drive is not in use.

A variety of engines were used in Series I and II vehicles but many early Land-Rovers are now fitted with later engines including V8 units.

and the whereabouts of some of these vehicles is known today.

The production models started to appear from July 1948. The specification included a welded steel box section chassis, the 1595cc engine of the Rover P3-60, a relatively simple body in Birmabright 2 (an aluminium alloy), and a canvas tilt. Until August 1950 all Land-Rovers had permanent four-wheel-drive, and a freewheel device was incorporated in the drive to the front wheels so that they were being driven only under acceleration. Later models still had four-wheel-drive but its selection was at the driver's discretion.

Series 1 Land-Rovers underwent a vast number of modifications but the principal milestones were the engine and wheelbase changes. The original 1595cc petrol engine was bored out to 1997cc late in 1951. In 1954 the original 80″ wheelbase was extended to 86″ and an extra model with a 107″ wheelbase became available. In 1955 the Rover P4 type engine was adopted with a cast iron cylinder head (instead of the aluminium cylinder head which was used on the cars) and in 1956 both models had their wheelbase extended by 2″ to 88″ and 109″ respectively. The reason for this seemed mysterious at the time but the extra 2″ was in the rear part of the engine compartment and was intended to accommodate the diesel engine which Rover had been developing for some time in response to demand. The new 2052cc diesel engine was introduced some months later.

The series 11 Land-Rover was introduced in 1958 with discreet changes to the body design – doors and front and rear wings were reshaped and sills were fitted to hide the chassis, the bonnet was also altered – and when the earlier petrol engines had been used up a new engine of 2286cc was also introduced in 1966.

What to look for

A big advantage to would-be purchasers of Land-Rovers is that body and chassis corrosion may not be extensive but it will be visi-

Land-Rover engine specifications (Series 1 & 2)

	1948-1951	1951-1955	1957 on	1958 on	*1955-1958
Fuel	Petrol	Petrol	Diesel	Petrol	Petrol
Bore	69.5mm	77.8mm	85.7mm	90.5mm	77.8mm
Stroke	105mm	105mm	88.9mm	88.9mm	105mm
Capacity	1595cc	1997cc	2052cc	2286cc	1997cc
Bhp @ rpm	50/55 @ 4000	52 @ 4000	52 @ 3500	52 @ 4000	52 @ 4000
Max torque (lbs/ft @ rpm)	80 @ 2000	101 @ 1500	87 @ 2000	101 @ 1500	101 @ 1500

Note that the Rover P4 type engine used from 1955 was a different engine from the previous 1997cc unit. Few parts are interchangeable between these engines. The earlier engine can be identified by the external oil pipes leading to and from the oil filter assembly.

Buying A Land-Rover

Early Series I models had the type of dashboard layout shown here...

...and this type of seating in which the rear of the seats were somewhat crudely sprung against the cab's rear bulkhead.

The dashboard layout on later models – here you can also see the controls for the four wheel drive and the transfer box...

ble. Body damage, if any, will be obvious, and although the rear end of the body is effectively in one piece the inner and outer front wings are bolt-on panels. Apart from some models which had steel bulkheads (which should be examined carefully) aluminium was used for virtually all of the bodywork and where aluminium comes into contact with steel the corrosive reaction which can occur in this situation has not been a serious problem.

Chassis and outrigger rust is another matter. The secret is to buy as good a chassis as

you can afford and bear in mind that if rust is visible in these areas it is probably more serious than it looks.

Look for evidence of tired road springs. They can flatten and even bow the opposite way. Swivel pins and track rod ends are also likely to be worn and should be checked. In general, mechanical components wear well and last a long time, but you are quite likely to find Land-Rovers which have remained in use long after extensive overhaul work has become due.

Very early models had a hydrastatic braking system for which parts are now scarce. If you come across one of these fairly rare models it would be worth bearing in mind that the later Girling hydraulic system can be fitted. The early Land-Rovers which were constantly in four-wheel-drive were fitted with Tracta constant velocity joints which are very expensive and sophisticated compared with the later Hardy Spicer universal joints.

Rover engines are fairly sturdy units but you should look out for the symptoms of engine wear. Worn bearings (especially on early engines which had by-pass lubrication systems), worn cam gear, etc, should be borne in mind – these four cylinder engines probably had to work hard at some time and Land-Rovers were often expected to do more than they were designed to do.

There are very few of the diesel engines about now, and as parts were not interchangeable with the petrol engines spares are also scarce. If the diesel engines had a particular weakness it was cylinder head damage due to overheating because of a low water level.

...and the improved seating of the later models (the majority of Series I Land-Rovers for sale are in need of seat repairs). Sliding windows are behind the seats in this 107" wheelbase cab-truck model.

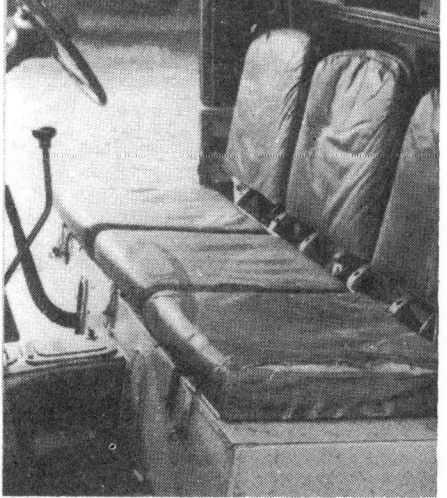

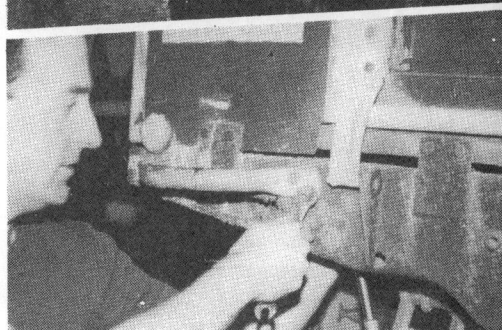

Weak points on the Land-Rover chassis are the centre rigger (top) and the front out-riggers (centre); the rear rigger (bottom) can also usually be expected to contain a lot of rot.

Which model to buy

If you are looking for one of the comparatively rare models, the desirable Land-Rovers will be the early 80" models, any of the 'Woody' station wagons bodied by Abbey Panels (but make sure it is genuine), also any Rover built station wagons because these are fairly rare too. □

MILITARY EXERCISE

Against all odds and far beyond the call of duty, 16-year-old Carl Brown single-handedly rescued this Land Rover in distress and restored it to pristine condition

Military vehicles and their restoration are a source of great fascination to an enormous number of people. Most seem to concentrate on the products of World War II but, presumably due to declining numbers of suitable vehicles, some of this dedicated band have turned to the post-war era for their raw material. Fully restored Champs, Humbers, Bedford RLs, Land Rovers and even Ferret armoured cars and Alvis Stalwarts are becoming more and more commonplace these days.

What many of the uninitiated may not realise is the incredible amount of work that these reborn military vehicles often demand. From the time of discovery in a scrap yard or old chicken shed to their first proud public appearance, many thousands of man hours will have been expended and more than a few grey hairs produced in the quest to bring the vehicle up to scratch.

Carl Brown decided that it was worth the risk of going grey when he set himself the task of completely restoring a 1955 Land Rover Series I FFR which had seen 28 years military service before being pensioned off. Mind you, he was only 16 at the time!

Carl has restored the vehicle to its original late 1955 divisional markings of the 3rd Armoured Division (black triangles on a red disc) which indicate that it was originally assigned to a signals detachment, being vehicle number 10 in the 3rd signals section (blue and white square markings) and fitted for radio (FFR) which ran off a separate 24-volt battery power supply. At some stage, approximately 1956, it was on overseas service in a desert location, hence the sand-coloured paint.

In 1964 the Land Rover was handed over to Abingdon Combined Cadet Force (CCF) and then sold on to Radley College CCF in 1970/71, before subsequently being sold to Bearwood College CCF. While at Bearwood, vandals got their hands on the Land Rover and seriously damaged it — so badly that it was just left to rot. Many components such as the dash panel, mirrors, gear knobs, seats and wipers were missing, the tailgate was twisted, all the lights were broken, the wiring loom had been ripped out and most of the engine components had been removed.

Carl, who was a Bearwood College CCF member, came to the rescue in 1983 when he bought the Land Rover with the intention of restoring it. As a 16-year-old, Land Rover-mad, student, he almost certainly didn't realise what he had let himself in for.

With enthusiasm overflowing, Carl started work immediately and slowly stripped what was left of the vehicle down to the chassis and running gear, ploughing his way through what seemed like 28 years of muck. After an initial clean up, the pile of bits left on the garage floor was steam cleaned but even that only removed about 80 per cent of the dirt. The remaining grime had to be physically scraped away with wire brushes and a great deal of elbow grease.

Once cleaned, the chassis was found to be in remarkably good condition for its age; in actual fact, the vehicle had only covered 28,000 miles since new. There were no bad rust patches and the only serious damage was to the offside of the rear crossmember which at some time must have been hit by another vehicle and crudely repaired. The whole chassis and both axles, once stripped of paint, were treated for rust and then primed, undercoated and finished off with two coats of gloss green (with engine and gearbox removed). Major items such as the fuel tank, propshafts etc were also cleaned and painted while new half-shaft bearings and oil seals were fitted to the rear axle.

With the engine and gearbox out of the chassis, they were stripped and cleaned and all missing items were replaced, before decoking the engine and regrinding the valves. The gearbox had to be cleaned out thoroughly when it was found to be full of glass and pine needles! New gear sticks and mounting brackets

Below left: **The Land Rover as found by Carl, after the vandals had done their worst.** *Below right:* **Having scraped away 28 years of grime, the chassis was found to be in remarkably good condition.** *Bottom left:* **A new canvas hood was fitted and all the original symbols were hand painted.** *Bottom right:* **The radios were in full working order once the correct parts were fitted**

were also added.

The braking system was non-existent and required total replacement — new brake pipes, master cylinder and linings. The exhaust system was also replaced, a new front bumper added and other minor details attended to before the chassis was complete.

The body was fairly straight with no major dents or creases but minor damage on the curves of the front offside wing and the rear tailgate. Once all the bodywork was removed, however, the bulkhead was found to be beyond the abilities of even the most skilled welder and it was obvious that a new one would have to be found. A great deal of searching was necessary before a bulkhead could be located in reasonable condition, which was supposed to be renovated but actually required considerable time and money to bring it to the required standard. The tailgate also had to be completely dismantled to allow the dents and twists to be removed, joints rewelded and, for strength, a new centre strip to be added.

During its 28 years of military service, the Land Rover had acquired approximately eight coats of thick paint which had to be stripped to bare metal to provide a good base for the fresh coat. All inaccessible points on the body were painted, such as the inside wing, bulkhead and under floor. Then the body was refitted to the chassis, replacing all bolts with new, liberally coated with 'copper slip' and all external surfaces were stripped. A minimal amount of body filler was applied as this does not key very well to aluminium.

Military markings exposed during stripping were carefully recorded for

future research into the vehicle's military history. Before painting, the body was allowed to dry and any joints and metal seams were thoroughly dried using a hairdrier. The body was painted using 'Tekaloid' industrial oil-based paint, two coats of primer undercoat (pre-mixed), followed by two top coats of bronze green, brush-painted (due to lack of funds). When dry it was polished with rubbing compound and treated to a damn good waxing.

To finish the vehicle, the wheels were painted, picking out correct colour wheel nuts; military markings were hand-painted with Humbrol enamel paint; new hoop sticks were fitted and a new canvas hood purchased. Other small details were tended to and finally the radios were installed, all fully working once the correct mountings, clamps and earthing straps were purchased.

The total cost of the restoration was approximately £1500 excluding the basic cost of the vehicle and radios. Fortunately tyres and several other items were serviceable and did not require replacing. The restoration took Carl in excess of 12 months to complete (including two house moves) but he's lost count of the hours of hard work involved. As can be expected when restoring an older vehicle, replacement, original pattern parts can often be difficult to come by, particularly such items as light lenses which are particular to military vehicles. Scrap yards proved to be a worthwhile source of good second hand (cheap) parts, and Carl is certainly convinced that all the searching was worthwhile. He didn't even turn grey, but now he can enjoy the pleasure of watching other military vehicle enthusiasts turn green with envy!

Choice

Looking for a utilitarian Classic? James Taylor
suggests the early Land-Rover as affordable fun

MOST people are familiar with the Land-Rover's origins in the late Forties as an agricultural/industrial vehicle intended to occupy vacant space in the Rover factories until a new car could be got into production, and with its makers' aim to export sufficient quantities to earn Rover larger steel allocations under the Government's scheme of the time. Most people know, too, that the Land-Rover was cribbed from the wartime Jeep, and that the first forty-eight pilot-build vehicles were made in 1947-48. After that, though, the vehicle's production history becomes rather complex.

Even that of the Series I models, the subject of this article, is far from straightforward. Until 1954, all Land-Rovers had an 80-inch wheelbase and a four-cylinder petrol engine, which until 1951 was of 1595cc but thereafter of 1997cc. Customers clamoured for greater load capacities, however, and from 1954 the 80-inch model was replaced by two new variants, one with an 86-inch wheelbase and the other with a 107-inch wheelbase, both still available only with the 2.0-litre petrol engine. Then for 1957, the 86-inch became an 88-inch, and the 107-inch became a 109-inch (with one exception, noted below). The extra two inches in the wheelbase of these new models allowed the fitting of a new direct-injection diesel engine as an option in place of the 2.0-litre petrol unit, which remained standard. Nevertheless, these were to be short-lived variants, for the Series II models arrived in 1958.

Body and interior

From the beginning, the basic Land-Rover body was an open pick-up. Early vehicles mostly had canvas tilts which also formed the cab roof, but it became possible later to buy a metal cab and metal hardtop for the load-bed, and the majority of vehicles sold in this country had at least the former. A large variety of adaptations within the basic framework of these bodies was available from the earliest days, but it would be pointless to detail them all here. The station-wagon bodies do deserve mention, however. Between 1948 and 1951, a total of 641 seven-seater station wagons were made on the 80-inch chassis, but the model was discontinued mainly because the cost of the coachbuilt bodies was so high and because, as passenger-carrying vehicles, these variants incurred Purchase Tax. Then in 1954 came a more spartan seven-seater on the 86-inch chassis, and a year later a 10/12-seater on the 107-inch chassis. The 86-inch became an 88-inch three years later, but there were no 109-inch Series I station wagons, the 107-inch model carrying on until the arrival of the Series II vehicles.

In fact, the body is probably the last thing to worry about when inspecting a potential Land-Rover purchase. Most of its panels are of aluminium alloy and do not rust, although they do dent fairly readily and a vehicle which has had a hard working life will probably bear the scars. Happily, it is not too difficult to secure replacement panels, unless of course the vehicle is an 80-inch station wagon, in which case replacement panels for the body (as distinct from front wings and bonnet) simply do not exist. These coachbuilt bodies, made by Mulliner's of Birmingham and Abbey Panels in Coventry, also had wooden frames, which by now are likely to have rotted. Replacement timbers are not available. The more ordinary Land-Rover bodies have frames and certain other parts of steel, and of course steel does rust. The commonest problem areas are all at the front of the body: in the footwells, at the base of the door pillars and around the door hinges, in the tops of the doors above the glass (except in early models with sidescreens), and in the frames at the door bottoms.

Inside, layout instrumentation changed over the years, and so did the shape of the seats. It is worth checking a potential purchase to ensure that all is well in these areas, as replacement parts of the correct type will be hard to find. Very early vehicles had seats with a nearly pointed backrest, while later cars had seats with flatter tops. Pre-1952 cars had tiny, almost unreadable, instruments, while the later models had two larger and clearer dials. In keeping with the generally spartan nature of the Land-Rover, a heater was only an optional extra, and the Smith's unit offered which bolts to the inner front bulkhead is crude in the extreme.

Chassis

The Land-Rover chassis is of hefty box-section, with the side-members welded together from sheets of steel plate. Extra rigidity is afforded by substantial box-section cross-members. Although it is extremely tough, it can rust, Rover's early resolve to galvanise the chassis of their off-road vehicle having fallen by the wayside before production began in earnest.

The drain holes provided in the underside of the box-sections can get blocked, and water trapped inside will set up corrosion. The whole of the chassis behind the rear axle is especially prone to rust thanks to its regular bombardment with mud thrown up by the rear wheels, and the rear cross-member merits careful inspection, as do the rear spring mountings on long-wheelbase models. Rust is commonly found on the chassis outriggers. Welding repairs will usually have been confined to the visible parts of the chassis, but the tops of the box-

sections can also rust through, and intending purchasers should remember that repairs here entail separating body and chassis.

Rough terrain can of course cause chassis damage, and the cross-members seem to be the most vulnerable areas for this. In severe cases, that under the gearbox may even be partially torn off, while the one under the clutch housing may be pushed right up against the housing itself and so cause engine vibrations to be felt right through the vehicle.

Steering, suspension

There is lots of scope for trouble in these areas! Some free play at the steering wheel rim is only to be expected, but there should not be much play in the drag link joints and track rod ends. In a car which has been roughly treated, the steering box or its relay might even have come adrift from their mountings!

Suspension is of course by leaf springs on all four wheels, with telescopic shock absorbers and rubber bump stops to check axle movement. Springs rust and can break, and it is worth checking that the ends of the shorter leaves are not wearing into the longer ones; if they are, imminent breakage can be expected. The other problem with springs is that there

are several different types: all are handed, there are heavy-duty varieties, and the front springs on diesel models are heavier than those on their petrol counterparts. Some owners may have fitted the wrong replacement springs, which can quite seriously affect a Land-Rover's handling.

The battering which a vehicle used over rough terrain can receive on its underside means that it is especially important to check the condition of the brake pipes and hoses, and a common problem is brake pipes hanging loose after rust has eaten through the anchor plates securing them to the chassis. The drums and operating cylinders are generally trouble-free, though the drum-type transmission brake (the Land-Rover's 'handbrake') can suffer from ratchet failure or oil leaks. The latter are usually easy to rectify with new seals and linings.

Transmission

The Land-Rover employs a four-wheel-drive system and an auxiliary transfer gearbox giving a second set of ratios for off-road work, although its main gearbox is essentially that of the contemporary Rover saloons. Before October 1950, four-wheel-drive was permanently engaged in both High and Low gear ranges, and there was a freewheel in the transmission to counteract torque wind-up between the axles. After that date, it became possible to select rear-wheel-drive only in High range, although four-wheel-drive was automatically re-engaged when Low was in use, and the freewheel was replaced by a dog-clutch which brought the advantage of permanently available engine braking on all four wheels. A small number of short wheelbase Land-Rovers with rear-wheel-drive only was built for the military in 1957-58.

An expensive problem to which Land-Rovers are prone is oil leaks from the front hub swivels. When new, these swivels — visible behind the wheels and emerging from a housing to which the brake backplate is fitted — are chromium-plated and smooth, but they rapidly become pitted. This pitting then causes damage to the oil seal, with consequent loss of lubricant. An optional extra, fitted as standard to military vehicles, were protective leather gaiters, which are of course worse than useless if they have split. Replacement hub swivels are very expensive, and the condition of

Top, the open cab version of the 80-inch chassis. Centre, the 80-inch with detachable metal hard-top. Above, a rare 80-inch Station wagon

Left, it might not be ideal for fast cruising but the Land-Rover is in its element on rough ground and it's really tough too

Choice

suspect ones can be checked by jacking up the relevant wheel and rocking it to discover any excessive free play.

Clunks and knocks in the driveline are quite normal, and can mask real faults. A check can be made for backlash in the differentials by attempting to twist the front and rear propshafts while the vehicle is parked out of gear with the handbrake off (and the wheels properly chocked!). Any more than a quarter-turn of free play points to problems. As for the gearbox and transfer box, neither is subject to any major weaknesses, although the rough use which they have to endure can provoke trouble. The earliest gearboxes had synchromesh only on third and top gears but from 1950 synchromesh was extended to cover second as well, and the third gear ratio was slightly raised. Nevertheless, the absence of synchromesh on second does not necessarily indicate that an early gearbox is fitted: the synchromesh is a common casualty! First gear on both types of box can get noisy, and even chip teeth, and a common fault is for the box to jump out of gear. The best way of detecting this condition is to drive the car hard in the intermediates and to 'blip' the throttle suddenly several times. As for clutches, trouble is readily detectable through slip or judder, and replacement is fortunately fairly easy.

Engines

Both the 1.6-litre and 2-litre petrol engines are essentially detuned versions of saloon car units, and like everything designed by the Rover Company in the late Forties, they were intended to survive indefinitely. Both were variants of the same basic design, which used overhead inlet valves and side exhaust valves in a wedge-shaped combustion chamber which was ideal for the low-octane petrol available in the post-War years and should mean that two-star will be adequate now. They often sound tappety and often use enormous quantities of oil, but neither condition need cause too much concern. The first simply calls for some rather fiddly adjustment (which is why it is often neglected) and the second most commonly indicates wear of the rubber sealing rings in the valve guides. A light knocking sound from the top end will probably mean the pad-type cam followers are worn, while in more serious cases the camshaft itself may be the culprit. Camshafts for

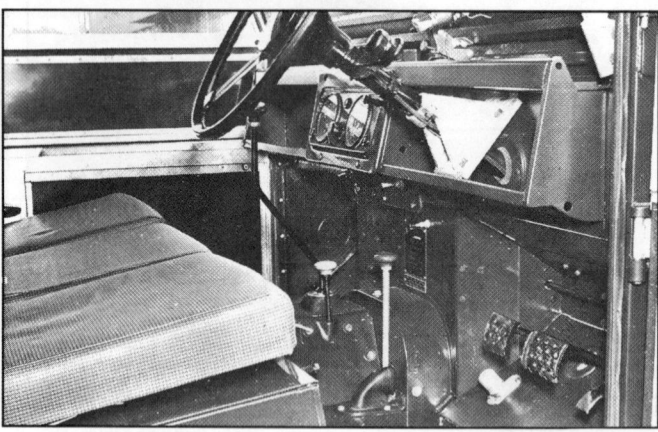

Above, the long-lasting diesel engine of the early Land-Rover. Left, front hub detail on a 1595cc 80-inch Land-Rover. Below, dashboard and controls of a 1952 version

the 1.6-litre engines are no longer available.

The 2-litre diesel engine is in fact neither as rugged nor as long-lived as its petrol cousins. It is noisy, even when in good order, and timing-chain clatter is very common. Smoke from the exhaust or excessive fumes from the crankcase breather will warn that a diesel engine is burning oil, which means that a top-end overhaul will be required. Like almost everything else on a diesel engine, this will be expensive. The diesel power units also sometimes crack their cylinder heads; this was, after all, Rover's first

diesel design, and the company did not *quite* get it right first time!

Purchase and maintenance

Although only a tiny proportion of the Land-Rovers for sale at any given moment will be Series Is, there is no real shortage of the beasts. The long-wheelbase models are harder to find than the short-wheelbase ones, however; 88-inch and 109-inch Series Is are hard to come by; and there are very few surviving 80-inch station wagons. Prices for the more uncommon variants will usually re-

flect their rarity, but it should be possible to pick up an 80-inch in reasonable condition for around the £500 mark. It is nice to report that not even really good ones are normally worth more than three times that amount, so that the purchase of a Series I Land-Rover should not entail a sheepish discussion with the Bank Manager. More than with most other 'Classics', it is worth considering *why* the vehicle is being bought. After all, there is no point in paying over the odds for pristine bodywork if the vehicle is going to be used regularly off the road. There are, incidentally, still several of the 48 pilot-build Land-Rovers unaccounted for, and it is not inconceivable that one may turn up somewhere despite the best efforts of enthusiasts to flush them out!

Parts for a Series I Land-Rover will not generally be available off-the-shelf from Land-Rover dealers, but there are many specialists who can help. These concerns regularly advertise in the specialist off-road magazines, but a few of the better-known ones are John Craddock, Ltd., of 60-76 North Street, Bridgtown, Cannock, Staffordshire (tel. 05435-77207); PA Blanchard & Co, of Foggathorpe, Selby, Yorkshire YO8 7PR (tel. 075-785-613); Dunsfold Land-Rovers Ltd., at Common House Road, Dunsfold, Surrey (tel. 048-649-567); and North Wales Land-Rover Spares, of 8 Vale Road, Rhyl, North Wales (tel. 0745-37623).

As for clubs, there is no shortage from which to choose. Most obviously, the Series I vehicles are catered for by the Land-Rover Series One Club, which can be contacted via David Bowyer, at East Foldhay, Zeal Monachorum, Crediton, Devon, EX17 6DH. This club aims to provide information and assistance to owners, and its services to members include technical advice and a spares location scheme, as well as a bi-monthly newsletter. The pre-1951 models (including pilot-build cars) are covered by the Land-Rover Register, which can be reached through Mrs. Sally Cooknell, at Langford Cottage, School Lane, Ladbroke, Leamington Spa, Warwickshire CV33 0BX. Finally, many Land-Rover enthusiasts have a special interest in off-road racing, for which the 80-inch models are specially favoured, and they are recommended to contact the All Wheel Drive Club, at PO Box 6, Fleet, Hampshire GU13 9YY, which runs a wide range of other events of interest to the owners of all types of off-road vehicles. ▲

TESTING TIMES

AN INTERESTING document has come our way. It's a copy of a final report dated August 27, 1949, carried out by the National Institute of Agricultural Engineering and Scottish Machinery Testing Station, on a test, requested by the manufacturers, of a prototype Land Rover.

THE TESTS span more than nine months of use of two vehicles, one right hand drive, the other left. The object was to obtain 'belt and drawbar' performance figures for the Land Rover and to ascertain its usefulness for general farm work and road haulage. The test was divided into three parts: Belt and Drawbar Performance; Road Trials and Farm and Field Trials.

The actual report consists of more than 30 pages of text, facts and figures, with graphs, which we couldn't possibly reproduce in full here. However, we are able to use some of the editorial and data which I think makes interesting reading.

For those of you who would like to buy hard bound copies of this particular report (No. RT 1/49034), they can be obtained from Tony Hutchings, Hillside, Harrow Lane, Steep, Petersfield, Hants. GU32 2BZ. Cost is £6.50 including p&p.

THE TEST

Brief specification:

The Land Rover is a light four wheel drive vehicle designed to transport goods and/or personnel by road or across country and for general use in farming operations. Two models were supplied for testing: one right hand drive (chassis No: R.32, engine No: 33) and one left hand drive (chassis No: L31, engine No: 32).

Engine:

Fuel:	Pool petrol
Arrangement and number of cylinders:	4 cylinders in line. 2.736in. bore × 4.134in. stroke. Aluminium alloy pistons
Compression ratio:	6.8:1
Engine speeds:	Governed 3000 rpm
Carburettor:	Solex 32 P.B.I. down draught including economiser and accelerator pump.
Air cleaner:	A.C. sphinx oil bath heavy duty type fitted with pre cleaner.
Ignition:	Lucas coil
Lubrication:	Forced feed from gear type oil pump delivering 160 gals/hour @ 2000 rpm
Cooling system:	Centrifugal pump, pressurised radiator and fan with cowling. Thermostat control. (Fine mesh radiator screen available as extra)
Electrical equip:	12v Lucas amp/hour lead acid battery. Headlamps, side lights, rear lights, horn, instrument panel, windscreen wiper, socket for trailer light, starter motor and dynamo.
Capacities:	Fuel — 10 gallons Cooling water—17 pints Oil—10 pints

The next section details chassis and body, vehicle dimensions, transmission, weight, drawbar heights, etc. Here is just a brief taster:

Overall dimensions of vehicle

On 6.00 × 16 tyres inflated to a pressure of 24 lbs/sq inch.

Length:	128.5 inches
Width:	62.5 inches
Height:	Without hood and windscreen folded down: 53.5 inches. Without hood and windscreen upright: 62.25 inches. With hood: 72 inches
Ground clearance:	To bottom of diff housing 7.5 inches. To bottom of spring clamp bolts: 7.75 inches

Transmission:

Clutch:	Single dry plate. 9 in. diameter. Foot operated.
Gearbox:	A normal gearbox giving four forward speeds and reverse (synchromesh on third and fourth) transmits power to a transfer box having two ratios — high and low. From the transfer box, output, two Hardy-Spicer open propellor shafts drive the front and rear axles: a free wheel on the front axle is incorporated in the transfer box.
Oil capacity:	Main box — 4 pints. Transfer box — 6 pints.
Wheels:	16 ins. rims fitted with Avon 6-ply 6.00 × 16 Traction or 7.00 × 16 Super-Traction tyres.
Wheelbase:	80 inches.

Moving on to the actual test. I will concentrate on the road and field trials only. There is extensive coverage in the full report, both in text and graphs of the Belt and Drawbar performance tests and the Farm Trials, where the vehicles haul anything from a two-farrow plough, a disc harrower, rolling and chain harrowing grasssland, to hauling a PTO driven muck-spreader!

Road Trials

These consisted of a number of long distance runs, including one hauling a trailer, during which fuel consumption was measured. All, except two of these test runs were carried out in the RHD vehicle. In addition to these journeys, both vehicles were used for transport of equipment and personnel on normal test work over a period of nine months, a record of mileage and fuel being kept. Wherever possible a Land Rover was used for hauling trailers and observations were made of its road worthiness and general behaviour. Maintenance of the Land Rovers was carried out by the Institute's garage staff and observations were made of the need for maintenance work over the nine month period. At the end of the test, the LHD vehicle was stripped down at the manufacturer's works and the component parts inspected for wear and damage.

Throughout the nine months of the NIAE tests, the vehicles were used as transport for everyday movements of equipment and personnel. They were found to be particularly useful for this type of work as they were capable of negotiationg rough land and being driven at speed on roads. It was also possible to tow fuel trailers and the tractor test dynamometer car (weight 3.5 tons).

The road journeys, both laden and unladen, were generally under 40 miles except for 10 trips of more than 100 miles. They were also used for crossing rough country, including hills, and farm lands, normally with a load in the body, but it is estimated that during the period, road work made up some 90 per cent of the total mileage. A record of mileage and fuel consumption was kept from which the following summary was obtained.

Again, space allows only snippets of information:

Right hand drive model
Total mileage: 13,800
Av. fuel consumption: 22.0 mpg

Left hand drive model
Total mileage: 9,240
Av. fuel consumption: 22.3 mpg
(After 5,000 miles, the performance tests and field trials started, and therefore the average fuel consumption figure given is taken over the period from 500-5,000 miles).

In addition, a number of specific test runs consisting entirely of road work were carried out, during which average speed and fuel consumption were measured. Mileage figures were checked with a map.

Driven at costant speed
Speed maintained at 30 mph (at no time exceeded). Unladen except for one passenger. Previous mileage 469.
The run was made largely on road route A1 at a time when free from heavy traffic and constant speed could be maintained. A stop was made at the end of each hour's run and the radiator water temperature was taken. The reading remained almost constant at 160°F.

Length of test:	158 miles
Av. speed	29 mph
Fuel consumption:	30.3 mpg.
No. of stops. 5 (each of 5 mins.)	

Driven at maximum speed
Unladen except for one passenger. Previous mileage 694.
The run was made during the return journey over approximately the same route as the first test. There was, however, goods traffic over some sections of the A1. One stop of short duration was made.

Length of test:	162 miles
Av. speed	42.9 mph
Fuel consumption:	23.6 mpg

The last four pages of the test report relate to comments and criticisms of vehicle construction and operation and conclusions. Again, I've reproduced only a fraction of the text available.

Comments and criticisms of details of construction and operation
1. Serious front wheel wobble was experienced on occasions with both Land Rovers at about 20-25 mph in the early days of the test. Modified steering relay boxes were fitted in order to cure this trouble which, however, re-occurred on the LHD model after ploughing trials and became progressively worse. No final decision has been reached as to the cause of this and it is understood that the manufacturers are continuing to investigate. Contributory causes were doubtless loose bolts clamping the steering box to the dash panel and worn tyres.

(Note: The manufacturers state that the number of steering box to dash panel clamping bolts has been increased from two to three and damping is now applied direct on to the steering swivel as in normal car practice).

2. On both vehicles, the sidescreen steady arms broke away and it was found that they could not be satisfactorily repaired by welding or brazing.

(Note: The manufacturers state that unsuitable steel was used in error and that these arms have been strengthened by an additional tie-rod).

3. It is suggested that both the engine oil and brake hydraulic accumulator filling points could be placed in a more accessible position.

(Note: The manufacturers state that the engine oil filler tube has been moved forward and the hydraulic fluid filler point is now adjacent to the fuel filler cap).

4. It was often noticed during the test that the rear body floor became hot owing to the proximity of the exhaust pipe beneath it. During the test a modified exhaust silencer was fitted and a guard placed over the portion of the exhaust pipe in the vicinity of the front nearside wheel to prevent objectionable smells due to burning of vegetable matter.

5. In the course of the test, it was found that the Land Rovers were capable of being driven almost anywhere.

There are another 15 specific questions and answers which make most interesting reading. The last part of the report deals with conclusions. Here are a few.

Conclusions
The results of the test show that the Land Rover is an excellent transport vehicle which can be used instead of a tractor for some agricultural operations, including belt work. It is capable of carrying a load of up to 10 cwt. across difficult conditions and of hauling trailer loads of the order of 4 tons.

On one test run an average of 42.9 mph was maintained over a 162 miles journey. At the higher speeds necessary to maintain such an average, however, the vehicle was too noisy for comfortable travelling.

Its capacity for draught work is shown by the results of the drawbar tests in which maximum sustained pulls obtained when unladen and equipped with 7.00 × 16 tyres, ranged from 2200 lbs on tarmac to 1700 lbs on rough, cultivated loam.

The results of the belt tests show that the Land Rover develops sufficient power for many farm operations such as driving threshing drums or medium sized hammer mills, although improvement to engine governing is desirable. The extensive and arduous nature of the tests to which the two models were submitted, showed them to be very robust vehicles capable of heavy work with low maintenance requirements.

It is suggested that modifications to the drawbar, gear ratios, driving position and parking brake system are particularly worthwhile.

Signed: T.C.D. Manby
(Officer in Charge)
M. Hamblin
(for the Director).
Dated: August 27 1949

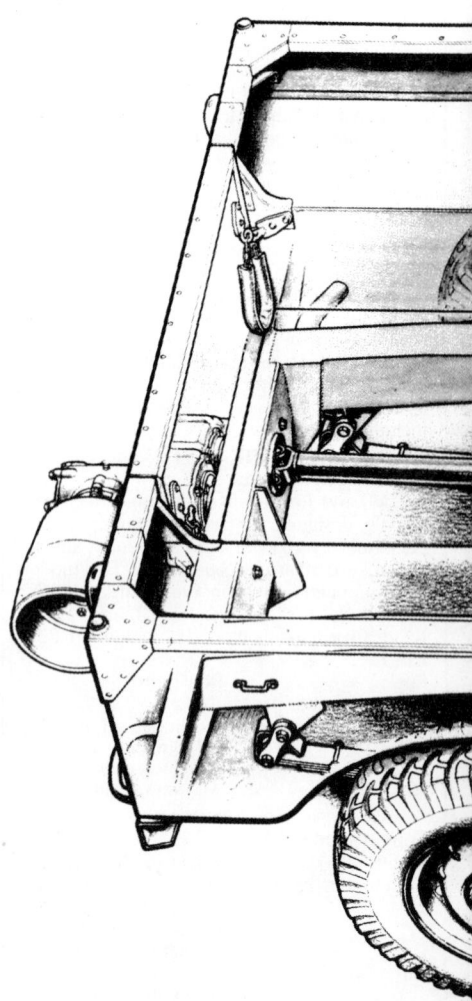

THE LAND ROVER AT 40

The British invention that changed the way the world thought about four-wheel drives is 40 years old this week. Bob Cooke looks back at how the Land Rover came about, drives one of the earliest and concludes that, as one of Britain's greatest exports, it deserves to be knighted

At the Amsterdam Motor Show on 30 April 1948 the covers came off a new British car intended as a stop-gap model, a temporary answer to postwar austerity. It was, said *The Autocar*, "a practical road and cross country vehicle built to high standards" and, 40 years on, not much has changed. The Land Rover is with us still and, after a career which has literally been felt in every corner of the world, shows every sign of its appeal being as durable as its shape and construction.

You could almost be excused for not realising that the car is 40 years old. Part of it may well be that the Land Rover does not seem to have changed all that much over the years,

The challenge was to design a successor to the Jeep using only components already available from the pre-war parts bin and then to make it sell worldwide. The box section chassis, for instance, was an economy move which paid dividends in the form of strength. The engine had to be an existing Rover car engine

but the main reason is that the Rover Car Company got the formula pretty well right — as much by accident as by design — first time.

Rover, renowned for its refined and elegant quality cars in pre-war years, found its plant turned over to defence production during the war. After the war, with nothing new in the way of a modern saloon car design to offer and a public somewhat short of car-buying cash, Rover had to come up with something to ensure its survival.

Even if there had been a new model ready to produce, Rover faced the problem of a steel shortage. It was director Maurice Wilks who came up with the answer: a British-built utility car bodied in aluminium — oddly enough there was no particular shortage of that expensive metal — with large-scale exports to developing countries in mind.

It has to be said that Wilks' utility started off on the right foot thanks to the existence of ▶

The Land Rover was put through its paces by armies the world over. The Belgian Army demanded three months of non-stop testing on tracks so tough that a new road had to be built to pull out the tanks that sank in the mud. Tests elsewhere involved a Kenyan safari and a dangerous, nine-week desert crossing

a very competent 'prototype'. This was no less a vehicle than the Willys Jeep, many of which had been pressed into agricultural service in Britain after the war to help overcome the shortage of tractors in an industry rapidly mechanising. Wilks' own estate was no exception, so Maurice and brother Spencer were well aware of the Jeep's value as an all-terrain vehicle.

The Rover design team bought two Jeeps from an army surplus dump and set about adapting Rover parts to serve in a similar vehicle, though one aimed purely at the civilian market. The similarity shows in the production Series One; not surprising, since the chassis of these first two Jeeps were used for the prototype Land Rovers. The Jeep's influence shows in the fact that the Series One has the same wheelbase, the same ground clearance, the same approach and departure angle.

The Jeep was a difficult enough target to achieve in the first place, since the specification was prepared by the US Army and manufacturers had to do their best to meet these strict requirements. In the event the final production car was something of a hybrid, mixing many Ford parts with bought-in components and the rugged 2.2-litre Continental engine from Willys, a low-down slogger of an engine perfectly suited for cross-country work.

Rover was not in a position to pick and choose its running gear and power plants. Lacking development money it was left with having to produce not just a feasible successor to the Jeep but a car that would conquer export markets using only components already available from the pre-war parts bin.

For instance, there was no question of tooling up for a professional-style chassis for the new car. The designers chose instead to weld up a box section chassis for each car from flat steel sheeting. A feature achieved by accident, perhaps, but it gave the vehicle excellent strength and rigidity and the current model's chassis is derived directly from those hand-made originals.

The engine, too, had to be a Rover car engine and, by the time early production models were ready for launching into a utility-hungry market, the 1595cc inlet-over-exhaust engine developed for use in the P3 60 Rover saloon was firmly in place. Another

accident, but once again ideal for a cross-country car. It produced only 50 bhp, but with a stroke of 105mm compared with a bore of 69.5mm it had excellent low down torque which peaked at 80lb ft at 2000 rpm.

The only complicated part Rover had to have specially made was the transfer gearbox. It had to supply drive to the front axle and also give the option of low-range gearing for work in difficult terrain. It took the latest generation of 90 and 110 Land Rovers to make the logical switch to permanent four-wheel drive. Novel? Not really — more a throwback to the car's origins. The very first Land Rovers had permanent four-wheel drive.

There was nothing as sophisticated as a central differential to get rid of transmission wind-up. Again, the Land Rover used a component standard in Rover saloons — the freewheel. Fixed between front and rear driveshafts, it allowed any tension to be released on the overrun. By 1950 this arrangement had been shed in favour of the simpler selectable system with dog-clutch engagement drive.

Transmission, too, a four-speed manual, came from the P3 saloon; so did the axles and springs. Only with the introduction of the coil-sprung 90 and 110 did the Land Rover's ride quality improve over that of the Series One, though the strength of the springing and competence of the damping had been refined over the years. As for the vagueness of that spindly 40-year-old gear lever, it too seems to have followed the design through the years, firming up at least a little with the introduction of the five-speed box.

There is no hurrying the Series One. Out on the road the low power of the engine means a cruising speed not much above 50mph and a down-change to cope with any hills. Off the road, it's the rigidity of the chassis and the simplicity of the suspension that limits the rate of progress across rough terrain and liberal use of the throttle can have occupants bouncing

about in their seats. Did someone say seats? There's no concession to comfort in this car — when Rover said utility it meant it.

Right from the start the Land Rover was a perfect exponent of the theory that the best way to drive cross-country is to keep progress slow and steady. The Series One car with its little engine, hand-made bodywork and sloppy gearchange, could still show many a modern all-terrain car the way across a stretch of tortuously twisting terrain. Select low ratio by easing the transfer lever into position; dab the lightweight clutch; select third, and leave it

there. Then get the engine up to around 2500rpm and leave it too, gingering the lightweight throttle just a little more when the terrain rises, just a little less when running downhill. Now the Series One will pick its way over almost anything with ease.

And there you have it — a cross-country classic. It is hardly surprising that within three years two Land Rovers were being sold for every one of Rover's saloon cars. And in the export market, demonstration of the prototypes generated more sales than the company could handle. ■

Land Rovers have been adapted for every kind of use including fire-fighting

Forty

The best of th

FORTY years ago, when those first early pre-produ
Land Rovers were being built, *Motor* magazine ed
staff took a vehicle for testing into the Welsh
around the Elan valley. The landscape was unsp
those days and has, remarkably, remained unsp
this day. So it was that the Series One Club decic
retrace the test route taken by the original
editorial team as much of it is still clearly definable
the illustrations published with the article.

As it turned out, a hundred and fifteen sh
examples of authentic Series Ones, made the t
Wales to celebrate the day the first Land Rover
on view at the Amsterdam Motor Show on Apr
1948. And the Dutch connection was still very
in evidence forty years later as a group of enthus
Series One owners from Holland were among th
to arrive in Wales.

Living as we do in a world that changes befor
very eyes, it was intriguing to stand in the very
in which the photographer stood forty years ag
clearly recognise every turn in the river and vir
every blade of grass from those early pictures. V
just goes to prove, perhaps, that although forty

●*Guy Pickford in pre-production R29 makes the hill climb that featured in **Motor's** 1948 report.*

●*Tony Hutchings (left) and David Bowyer (cent magazine's test Land Rover on the hill climb.*

●*Left: Parked up for the night at the Victoria Wells c a Dutch visitor.*

years on....
...but nothing's changed

ne's meet again on a Welsh hillside

eem like a long time to us, it's no time at all in of natural change.

nt organisers, David Bowyer and Tony Hutchings vided the route into four stages so that, with the ed plus entrants also grouped into four lots, the dual convoys would at least be manageable. It's s fun to cruise in convoy, especially when the ring turns heads and you get waves and smiles passers by.

four stages took in the magnificent Elan Valley oir complex, which feeds water to the Midlands; mains of the Dolaucothi Gold Mines at Pumpsaint; p and trecherous hill climb at Brechfa and the ful Towy Forest. So, not only did the gathering the chance to see some rare Land Rover nery, it also offered an opportunity to see some nd often spectacular Welsh scenery in the area d Llandovery.

h an excellent evening meeting place at the a Wells holiday complex, the entire weekend was nely well planned, and the Series One Club sers are to be congratulated on their enterprise. done lads, we all had a good time.

● *Voted best overall was Kenneth Wheelright's 437 DEL, with it's twin 438 DEL, both in perfect condition. Alongside is David Bowyers 86 inch Station Wagon.*

ocal farmer who, as a child, rode in the **Motor**
the hats!).

eth Poole's 56 inch Station Wagon alongside

● *Alongside R29 at the Gold Mine car park is LRO's very own 1949 80 inch HER 154.*

More pictures over the page

▲ *Leaving the Dolaucothi Gold mines.*

▼*This picture, taken on April 30, 1988 is almost exactly the same as the one that appeared in **Motor** Magazine in 1948.*

●*Above right: Dunsfold Motor Museum's 107 inch station wagon.*

●*Right: What can you say? Two Dutchmen in a kind of Series One!*

●*Below: We got ourselves a convoy. A stream of Land Rovers pours down the narrow roads of the Elan Valley.*

●*Best overall were Kenneth Wheelright's two pristine 49ers. David Bowyer (right) makes the presentations, Tony Hutchings looks on.*

INTERESTED IN LAND ROVER AND RANGE ROVERS ?

Why not join a club and meet others with similar interests ? There are Land Rover and Range Rover clubs all over the country and most of them belong to the ARC - the Association of Rover Clubs.

You can contact the secretaries below for more information:

THE ASSOCIATION OF ROVER CLUBS LTD
Andrew Stovordale, 65 Longmead Avenue, Hazel Grove, Stockport, SK7 5PJ
Tel: 061 456 8224

LAND ROVER SERIES ONE CLUB
David Bowyer, East Foldhay, Zeal Monachorum, Crediton, Devon EX17 6DH
Tel: 0363 82666 (business hours only)

LAND ROVER SERIES 11 AND 11A AND FORWARD CONTROL CLUB
PO Box 1609, Yatton, Bristol, BS29 4QP

RANGE ROVER REGISTER LTD
Chris Tomley, Cwm Cochen, Bettws, Newtown, Powys, SY16 3LQ
Tel: 0686 650430

ALL WHEEL DRIVE CLUB
David Sarsfield-Hall, Flat 6, 85 Henley Road, Caversham, Reading, RG4 0DS
Tel: 0734 483092

You can usually find a complete list of all the clubs throughout the country in Land Rover Owner Magazine - a monthly magazine available from your local newsagent.

LAND ROVER OWNER MAGAZINE
The Hollies, Botesdale, Diss, Norfolk IP22 1BZ
Tel: 0379 890056

R-Series Land-Rovers

AFTER the few jeep-chassis centre-steering prototype Land-Rovers of 1947, a batch of 25 (later increased to 50) preproduction prototypes was constructed for testing and evaluation.

Conceived as an all-purpose 'farmers friend' with a multitude of power attachments, the Land-Rover was intended to be a stop-gap by Rover until the business could get back into full swing with car production after the war. It was announced at the end of April, 1948 at the Amsterdam Motor Show where two vehicles were shown. They were well received and production started in July, 1948 with chassis numbering from 860001.

From the original 50 sets of parts only 48 vehicles were constructed, numbered from R1 to R48, and these differed in many respects from the later production vehicles. The R-Series all had galvanised chassis, the front bumpers were integral with the chassis and the steering column support bracket was made from three pieces of welded up scrap metal! Seat backs were rather rudimentary curved pads. The upper door panels, in other words the side windows, had no hinges and were rather ineffective, and the door panels were not reinforced at the hinges. The door handles were in the form of 'gate-latches'.

I first became interested in these early vehicles when I saw R14 advertised in 1974. Later I acquired the remains of R29 which I restored during 1974-75. A lot was learnt about early Land-Rovers from this rebuild. R29 was shown at the National Rally at Eastnor and the Leyland-Donington Rally during 1976 where many people were most helpful in pointing out minor faults!

It was following a write-up in the

Tony Hutchings, who runs the Land-Rover Register has restored two of the rare pilot-built vehicles like R4 above – and is looking for more

motoring press that R4 was offered to me. This vehicle had only had three owners from new and had spent the first eight years of its life four miles down the road. Subsequently it went to its third owner miles away in Sussex in 1956 where it remained untaxed on a small-holding. The asking price was the price when new!

When purchased the Land-Rover was complete but in a sad condition with a very rusty scuttle which required partly rebuilding in new materials. The first operation was totally to dismantle the vehicle and store the pieces in an empty garage. The integral bumper ends had broken off which was a common failure of all the early R-Series. The Skilcentre in Portsmouth came to my rescue and fabricated a new bumper which the local blacksmith in Petersfield, Steve Pibworth, kindly welded on for me during a half day spent at his forge. Various small imperfections in the chassis were corrected then. The next step was to regalvanise the chassis as we had disturbed the zinc in welding. A local shipyard re-dipped it. Originally the chassis had to be dipped twice which resulted in some twisting of the members and was probably the reason why the production vehicles were only painted.

The next step was to overhaul the 1595cc P3 engine, which was an early replacement engine installed when Rover sold off R4 in December 1948. Incidently, when those early vehicles were sold they were up-dated with later door handles,

seats etc. David Moss of Petersfield undertook this work for me, with Michael Hodgson (who had previously worked with me on R29) sorting out the axles which have non-adjustable Girling brakes. I sorted out the gearbox which, in these early R-Series and the first 1500 or so of the 860000 series, has a car-type free wheel as the vehicle has permanent four-wheel drive.

A modern wiring loom was acquired, and after the Skilcentre had again come to my rescue with the necessary parts for the scuttle and doors, I started on the body. The aluminium panels were of a very heavy gauge which along with the galvanised chassis, probably helped to prolong the life of these early vehicles. The back corners were particularly bad though and I had to compromise and add later-type corner pieces. The body is basically in three parts, two of which – the rear and centre sections – being easily removed in minutes. Peter Redman of Petersfield made the correct pattern seats and backs as he had done before on R29.

The finished colour is a rather pleasant grey-green which was standard until the later bronze green came in. A short canvas cab has now been fitted which was the only protection available in early 1948. The wheels are of the split-rim type fitted with Dunlop Trakgrip tyres, The rebuild took just over six months and was carried out for a very modest sum.

My researches have now accounted for eleven of the 48 R-Series: R1,3,4,14,19,24,25,29,32,45 and 46 which exist in various states of rebuild or decay. R1 is on show at the Leyland Historic Collection at Donington and my own R4 is on show at the National Motor Museum at Beaulieu. □

LAND LORDS

The Land-Rover might be the boss off-roader, but just how practical a classic choice is Solihull's venerable four-by-four? Bob Cooke takes a look at the original Series I cars. Photos: Julian Mackie

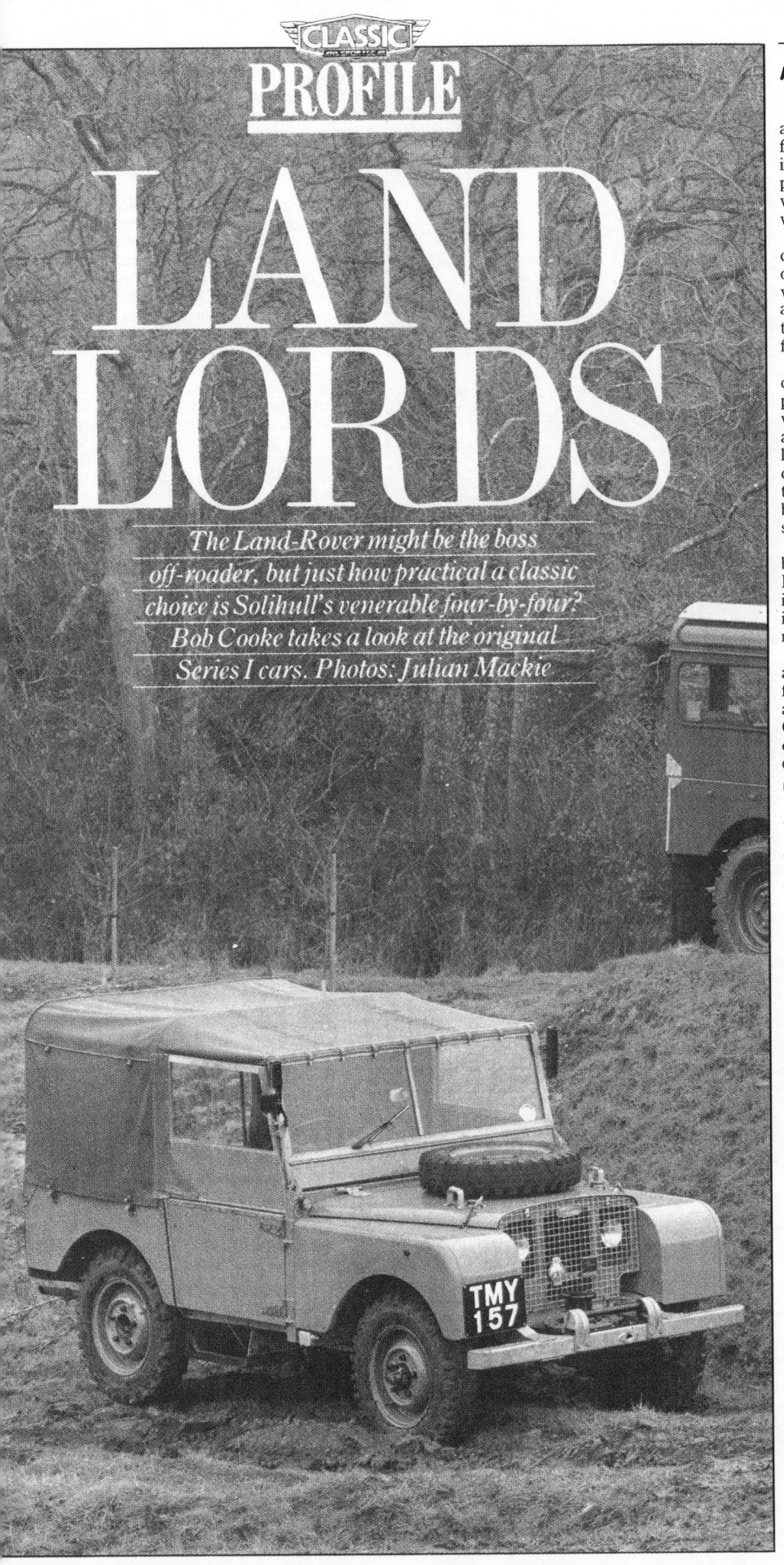

There are two levels of rarity surrounding Series I Land-Rovers. By far the most exciting, interesting and therefore most collectable models are those that appeared during the first four years or so; it's not merely that these hold innate 'first production model' value – which they patently do, in any case – but also because Rover was still experimenting with specialised versions, of which very few were built.

A typical example is the 80in station wagon, which could be seen as an early attempt to lift what had originally been conceived as a purely agricultural vehicle into the realms of real passenger cars; another would be the early military version which, to satisfy the demands of the army chiefs, had to be fitted with a Rolls-Royce engine.

The cachet of owning one of these early specialised cars matches that of owning one of the original 48 pilot-build models. The problem facing the enthusiast wishing to acquire one of these is that the whereabouts of existing surviving examples is well catalogued by the various Land-Rover enthusiasts' clubs, and as the result of rapidly growing interest in the marque over the past 10 years, almost every barn, farmyard and junk yard in Britain has been scoured for signs of any others.

There remains, however, a second level of rarity, based on the fact that most of those early Land-Rovers were specified to individual orders – as indeed many present models are today, and so interesting minor variations will exist from model to model.

The most exciting models of all are, unfortunately, well out of the grasp of any enthusiast – the first prototypes that set the pattern for what began as a stop-gap measure to see the Rover Car Company through a few tough post-war years, and ended up as one of the most far-reaching automotive developments of all time.

The whereabouts of existing examples is well catalogued by Land-Rover enthusiasts

Rover, stuck with a badly-bombed factory at its normal Coventry home and with work dying out at the Air Ministry factory it had been running in Solihull, needed a new product to get itself back into the car-making business. All it had to offer was a revised version of its pre-war saloon, which it appreciated was outmoded and too expensive for a straitened, war-ravaged economy. It toyed with the idea of switching from the traditional Rover concept of providing elegant luxury cars for the well-to-do to build a base, economy-minded miniature 699cc car, code-named M1, not unlike the Morris Minor in concept. Fortunately for Rover, Government policy dictated that allocations of precious steel favoured companies that earned good export orders; Rover had traditionally limited its production to home market requirements, but now interest turned to the world at large. When it was realised that world demand in fact favoured larger cars, the M1 project was dropped and attention focused on the P3 60 model.

It was pure chance that Rover's chief engineer at the time, Maurice Wilks, had been using an American Army-style Jeep on his Anglesey estate, where it gave stout service as part-time tractor and general runabout. Casual talk with brother Spencer – chairman of Rover – turned one day to the subject of what vehicle would replace the Jeep when it finally wore out. There and then, so the story goes, the idea of the Land-Rover was born, spurred on by the obvious international demand for low-cost agricultural vehicles both from newly-developing nations and war-ravaged countries struggling to rebuild their agricultural infrastructures.

Rover engineers were set to work with two army surplus Jeeps as models from which to design a

quick-and-easy to assemble agricultural four-wheel drive vehicle, which right from the start was dubbed the Land-Rover. It was not meant to be a long running project, just enough to establish a good export record for Rover and to get some cashflow going to help with the development of more traditional Rover cars.

The first bodies were knocked up in aluminium alloy sheeting, not because this material was cheap but because it was readily available in the immediate post-war years while steel was still strictly rationed. Rover didn't mind the extra expense of the material since it could easily be hand-worked, which meant it didn't have to go to the high cost of tooling-up for mechanical manufacture of what was envisaged as a relatively short-run product.

The first prototype appeared in 1947 and was, in fact, assembled on the chassis of one of those army surplus Jeeps. It was powered by the 1389cc, 48bhp engine from the Rover 10, driving through a production Rover gearbox, and using Rover 10 back axle and springs. The only important specially-built item was the dual-range transfer box. In all, three or four prototypes were built, and during their development period they were tried with all the accessories like winches and power-take-offs that would characterise the later production jack-of-all-trade Land Rovers. One more feature was intended to suit the car's role as an international seller: the steering wheel was placed centrally, to bypass the left-hand/right-hand drive problem, leaving the driver – tractor fashion – to straddle the transmission tunnel.

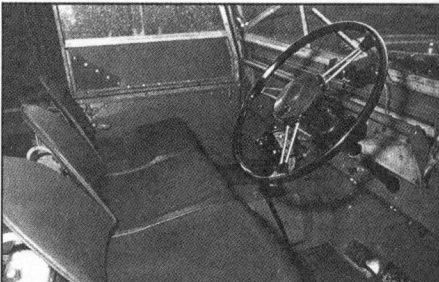

Top left: 86in station wagon put through paces – ample axle articulation helps car's off-road ability. Left: Early 80in – note the behind grille lamps. Above: Spartan but practical interior. Below: 1595cc engine in the 80in. Bottom: Handy period front-mounted winch

Sadly none of these prototypes survive. They did, however, serve to demonstrate the soundness of the Wilks brothers' decision to follow this go-anywhere trail, and 48 pilot-build models were turned out for further development and to be used as demonstrators. There was urgency in the air; to secure foreign orders the Land-Rover had to be shown internationally, and quickly, so the target was the Amsterdam Show in April 1948. Happily, the central steering wheel idea had been discarded; the pilot-build cars were all based on an 80in wheelbase (not surprisingly, that of the Jeep from which the Land-Rover had sprung) but some were right-hand drive, some left-hand drive. Most had simple open bodies but there were also estate versions.

The pilot-build Land-Rovers may not have stopped the Amsterdam show, but they did make an immediate international impact – so much so that within two years production Land-Rovers were outselling Rover's prestigious new P3 saloons by two to one.

Eighteen of those 48 pre-production models are known to exist to this day; No 1 has been bought back by Land Rover and restored, while No 4 is on display in the National Motor Museum at Beaulieu. Nine are known to have gone overseas – which leaves 21 others still to be located. Enthusiasts accept that many of them might well have found their way into breakers' yards to be irretrievably dismantled, but some are thought to exist, hidden away in forgotten farmyard corners or in desolate industrial wastelands, just waiting to be dragged out and reconditioned.

There was urgency in the air...
The Land-Rover had to be shown
internationally, and quickly

Most desirable of the production cars that followed are without doubt the first 1500, which varied only slightly from the pilot-build cars, even though this means they are the most basic of the early models. Purist collectors see them as the source from which all other Land-Rovers sprang, and therefore of particular historic value; again, not many of these are to be seen in Britain, but that's not hard to understand since even from the earliest days some 80 per cent of Land-Rover production was for export, and most of those early production cars will have ended their days in the depths of African wildernesses.

Also of interest are examples of the first 40,000-odd that made up the first three years' output, since these featured 'permanent' four-wheel drive; permanent in the sense that drive to the front axle was via a freewheel system, as used on Rover saloon cars. It did away with any need for the driver to worry about when four-wheel drive should be engaged or disengaged, yet it prevented the problem known as transmission wind-up – tension that develops in transmission of four-wheel-drive vehicles when wheels that are geared together turn at slightly different rates – as when negotiating a curve. Modern Land-Rovers have permanent four-wheel-drive, with expensive differential gears allowing all four wheels to turn at their own rates; the early Rover freewheel did the same thing for the first Series I cars by transmitting drive to the front axle only under acceleration, leaving them to freewheel on the overrun. It was an elegant solution, but not without its problems, mainly that braking under engine compression, an important requirement in serious off-road driving on slippery surfaces, was less effective in these models, since only the rear wheels were able to handle that braking force. It was partly that problem, plus a continuing search for overall simplicity, that led to the introduction of a dog clutch arrangement that allowed switching from four- to two-wheel drive.

Most of these early models were open pick-ups, but there were some 640 seven-seater Station Wagons, the first examples seeing the light of day at the 1948 Earl's Court Motor Show. This country estate, a concept that foreshadowed the appearance of the Range Rover, featured a curvaceous alloy body fitted to a wooden frame, and refinements included a single-piece windscreen and real wind-up windows in the doors. The extra manufacturing work resulted in too high a price penalty, though, and the seven-seater configuration meant that the car was patently not a commercial vehicle and therefore attracted Purchase Tax. In a country still short of Green Wellington money the Station Wagon was turned out at a rate of under 200 a year and eventually went out of production in 1950. Few appear to have survived; again most were exported, and only 18 are known to British enthusiasts, naturally jealously guarded by their proud owners.

Though the army was busy commissioning a four-wheel-drive vehicle of its own – the project that would turn into the Austin Champ – it was naturally interested to see how the Land-Rover would behave. It was impressed with the car's capabilities, to the extent that some 150 were ordered. The army wasn't sure that the lightweight 1595cc engine was gutsy enough, and felt the 2.8-litre Rolls-Royce B40 specified for the Champ would be better; it also liked the idea of standardising engines, even though the Champ and Land-Rover were seen as fulfilling different roles – the purpose-built Austin as a front-line machine and the Land-Rover as a behind-the-lines support vehicle. Accordingly, 33 Series I 80in

Above: Most desirable Series 1? The very rare 80in Station Wagon. Below: First known use of Land-Rover for a ceremonial review, 1951 – George VI and Field-Marshal Alexander

cars were re-engined to suit such a purpose.

'Re-engined' is an over-simplification; re-engineered might be better, since the much greater weight of the big Rolls block meant heavier gauge springs, bigger clutch and different brakes for the little Land-Rover. It meant, too, a small stretch in the wheelbase to accommodate the larger engine, with the result that the 33 Rolls-engined Land-Rovers are also known as the 81in models.

One of these 1949-year models was entered in a series of military trials for comparison against the newly-produced Champ. Rover entered one of its own basic 1595cc models. The standard Land-Rover comfortably outperformed everything else, including the purpose-built Champ. The 81in Rolls-engined models were taken out of service within a year. Only six of these have been traced by present-day enthusiasts.

In a country still short of Green
Wellington money the Station Wagon
was built at the rate of 200 a year

General appearance of base Land-Rovers changed little in the 10 years of Series I production. Points to look for are the lack of exterior door handles on early models, which also had the front sidelights mounted on the windscreen bulkhead, and headlamps completely covered by the wire mesh radiator guard. The headlamps appeared through the mesh in 1950, the sidelights moved to their more familiar position on the wings in 1951, and, also in 1951, external door handles made their convenient appearance. At about the same time an enclosed cab and rear hardtop became available.

The 1595cc engine lasted until 1951, at which point increasing customer demand for a vehicle with better slogging power forced Rover to bore it out to 1997cc. The power increase was a minimal 2bhp, up to 52bhp at 4000rpm, but torque soared from 80lb ft at 2000rpm to a whopping 101lb ft at a lowly 1500rpm, resulting in a powerplant far better suited to towing and for driving static machinery. A search for bigger load capacity resulted in a wheelbase stretch for the basic model, up to 86in, plus the introduction of a long-wheelbase 107in version. The opportunity was taken at this point to recess the door handles, a feature that has become familiar on all Land-Rovers right through to modern 90 and 110 models.

The number of body style options had blossomed by now, too, with the basic pick-up model joined by

chassis-cab, canvas-tilt and, once again, Station Wagon versions of both 86 and 107 types. This time round, though, the Station Wagons had simple panelled bodies on light metal frames, much cheaper to produce than the earlier coachbuilt models, and therefore much more accessible to a marketplace that was also growing in wealth. Already in these early estates could be seen the makings of the modern series of classy County-trim Land-Rovers.

These two developments of bigger engine and longer wheelbase were in fact unrelated, each answering in their own way a demand from customers seeking more power in vehicles required to carry ever heavier cargoes. When in 1956 a 2in wheelbase stretch occurred, this had nothing whatever to do with demands for greater load capacity. The 2in stretch all went into the engine bay, and was purely to accommodate a new alternative powerplant, the 51bhp 2-litre Rover diesel, even though this unit did not in fact make an appearance until the following year.

That was just a year before the Series I officially came to an end. In celebration of 10 years of exciting development, Rover produced a smoother, smarter range of cars, taking its temporary post-war stopgap project to new heights and into lucrative new export markets.

Total production of Series I cars falls not far short of a quarter of a million. Many are still in service, a tribute to the basic strength and soundness of the simple but competent design and construction. Naturally many didn't last the pace of 40-odd years of hard work, and have found rest in scrapyards and in the back corner of many a British barn.

BUYER'S SPOT CHECK

Body: It's only partly true that because Land-Rovers are built of aluminium they don't corrode; there are many steel plates and brackets that *do* rust, and there can sometimes be blistering at steel-aluminium joints. Land-Rovers that have been used on farms often had to carry sacks of chemical fertiliser which in some cases did react with the aluminium.

Nor is corrosion the only problem; because Land-Rovers were basic, agricultural or industrial workhorses they were often treated in a very rough and ready manner, with the result that dents and creases in the bodywork could be quite common. There's good and bad news; the bad is that original replacement panels are unobtainable, and aluminium isn't the easiest metal to work on, since it can't be welded or brazed in the usual way – and being rather brittle may not respond well even to simple panel-beating treatment. The good news is that because Land-Rovers are built on a strong, rigid separate chassis, holes or rotten patches of bodywork won't affect the overall structural strength of the vehicle. It is also quite a simple matter to replace bodywork with replica panels, though obviously if you have to pay a specialised metalworker to make the parts you'll be paying a heavy premium.

Chassis: Even a car sporting a seemingly pristine bodyshell could be a complete disaster underneath. Only the pre-production 48 models had the ideal of galvanised chassis frames; production cars relied purely on the innate thickness of heavy-gauge steel to ensure a satisfactorily long service life. Remember, after all, that the Land-Rover project was not originally conceived as a long-term operation. By far the most important examination should be of that chassis frame. Remember that Land-Rovers will have been driven over rough terrain, and the chassis itself could have taken quite a pounding – particularly the front bumper and the central cross-member (1) which passes under the transmission: it protects the transmission, but in warding off those bumps and scratches naturally opened itself to serious corrosion problems. Side extensions from the chassis holding the body panels (2) can also suffer serious

Ten things you didn't know about Land-Rover Series I

1. The first Land-Rovers had permanent four-wheel drive, a feature deleted in 1950 and not reintroduced until the appearance of the coil-sprung 110 model in 1983.

2. Foreign build of 80in Land-Rovers began as early as 1952, when Minerva of Belgium fitted British-made 1997cc engines and running gear to hand-made Belgian chassis and bodies. Production stopped in 1956, but many Minerva Land-Rovers are still in use by the Belgian military services. Though agreement was reached for Santana of Spain to build Land-Rovers in 1956, no Series I cars were made there; start of production was delayed to 1958 with simultaneous introduction of Series II cars in Britain and Spain.

3. First Land-Rovers were available in any colour you liked as long as it was Sage Green. Only in 1949 did the more familiar dark green appear, and it was not until 1954 when the 86in and long wheelbase 107in models appeared that a choice of grey or blue was offered as an alternative to the dark green.

4. There were two distinctly different 1997cc engine blocks used in Series I cars; both had 77.8mm bore and 105mm stroke, both had a 6.8:1 compression ratio, both had three-bearing crankshafts. The first unit, fitted from late 1951 until late 1953, was a bored-out version of the 1595cc Rover P3 engine; after 1953 this was replaced by a version of the P4 60 saloon engine which had identical cylinder dimensions. The difference? The earlier engine had siamesed bores, and threatened to suffer from bore-scuffing and overheating. The later one had equally-spaced bores with water flow between all cylinders, overcoming these problems.

5. The first half-year's production of Land-Rovers used drivetrain components from the 10, 12 and 14hp pre-war Rovers which went back into limited production after the war. When in mid-1948 P3 production started, the Land-Rover had to take the new saloon's higher-ratio final drive, thus reducing its steep-slope climbing ability; later it had also to take the P4 gearbox with its higher third gear.

6. The first diesel-engined Land-Rovers had supercharged, two-litre two-cylinder engines made by Turner of Wolverhampton. These, along with a 3-litre version, were offered as a conversion in 1953. In 1955 'Operation Enterprise' was staged to prove the engines' durability; this involved driving a Turner-engined 107in pick-up from Wolverhampton to Nairobi across the Sahara desert. During 195. Rover considered fitting a tractor diesel engine, but eventually decided to make its own. The Rover diesel was a technically advanced unit of 2 litre capacity with overhead valves, wet-liner cylinders, Ricardo Comet combustion chambers and aluminium alloy pistons. Power was an impressive 51bhp, torque 87lb ft at 2000rpm.

7. Modern car-makers are using ever-more galvanised steel in their products to reduce corrosion. Land-Rovers, likely to be used in slushy conditions as a matter of course, naturally had galvanised chassis frames… or rather, the 48 pre-production models did. Ever since then, claims that Land-Rovers resist corrosion apply only to the non-rusting aluminium bodies, not to the chassis. These are prone to corrosion – but their very thickness and solidity is enough to ensure 20-odd years of life.

8. The gearshift lever on the first 1500 production cars was, as on the pre-production 48, mounted to the body – with the result that whenever the car bounced over rough terrain the gearlever was pulled out of the gearbox housing.

9. Land-Rovers were always meant to be very basic vehicles. The original intention was to offer doors, a canvas roof for the cab, a spare wheel carrier – and even the starting handle, spare tyre and passenger seat – as extra-cost options. In the event these were all included in the standard specification at an overall price of £450.

10. Rover started in-house conversions in the first year of production, offering first a Welder, an 80in with an arc generator driven from the power take off, and compartments for oxy-acetylene equipment. Neither this, nor the 1952 fire engine based on an 86in, sold well enough to warrant continued production, so Rover farmed out conversion work to specialist firms. Only in the past few years has Land Rover decided to take such work back in-house.

Above: One of the 33 experimental Army 81in cars, fitted with the Rolls-Royce engine. Below: 86in Station Wagon converted to a shooting brake for use of Duke of Edinburgh

rust, since these were in the front-line of water splashed up by the knobbly tyres. You're unlikely to have to use a screwdriver to poke through areas you suspect of being rusty – if the corrosion is there it'll be very visible, often with holes in the chassis members that you could get a hand through.

Look out for steel plates that have been welded over corroded patches. It's not necessarily a bad sign, because that was a common way of extending the life of a Land-Rover chassis, and a well-executed repair could almost match the strength of the original. Patched-up chassis looking rather like botched Meccano construction are not unknown on Land-Rovers still in hard-working service; remember that the very basic structure of the car lent itself to that sort of agricultural repair.

Certainly be very wary of a chassis showing obvious corrosion unless you are prepared to go to the expense of a complete replacement chassis. These are available – but they'll cost you, because they've been specially hand-crafted.

Otherwise the car's simplicity is enough to reduce the general structural repair bill – with flat glass windscreens, big bolt-on hinges and simple equipment, there's very little to go wrong.

Suspension, steering, brakes: There's much more to worry about here, again because these cars will mostly have led rough-and-ready lives, with many having been regularly overloaded, driven over unmade terrain and quite possibly denied regular servicing. Look out for sagging springs (3) and badly worn or corroded spring hangers (4) and shackles. Spring bushes are relatively simple to replace but do

Left: 1947 prototype, showing the central steering wheel. Right: West German border police, 1953, with Land-Rover built under licence by German firm Tempo. Below: 107in Station Wagon, with snow-clearing gear

be aware that many different types of spring were available for various Land-Rover models, and a mismatch could affect the car's handling.

The brakes are unlikely to cause a problem, but do check that brake piping has not been thumped into exposed positions as the result of snagging on rocks or stumps.

For the same reason the steering linkages might exhibit too much play. Expect to find some free movement in the steering wheel, but beware that excessive sloppiness could be caused by wear in the track rod ends (5), or may even be caused by the steering box (6) working loose from its mountings.

A more serious problem is the potential wear in the front hub swivels (7), a known weakness. Desirable as the earlier permanent-four-wheel-drive versions are, it's important to remember that the type of CV joint used in these models will be very expensive to replace compared with the more common Hardy Spicer universal joint used in later models.

Engine and gearbox: Not a lot goes wrong with these robust little powerplants, which are in effect retuned and strengthened versions of Rover saloon engines. Early engines might show signs of bearing and cam-gear wear, but these need not mean a rebuild if mileage on the restored car is to be kept low and slow. Excessive tappetty noise is probably just that and can be cured by normal adjustment, and excessive oil consumption – occasionally manifested in the form of a very smoky exhaust – is probably due to nothing more than worn valve stem rubber seals.

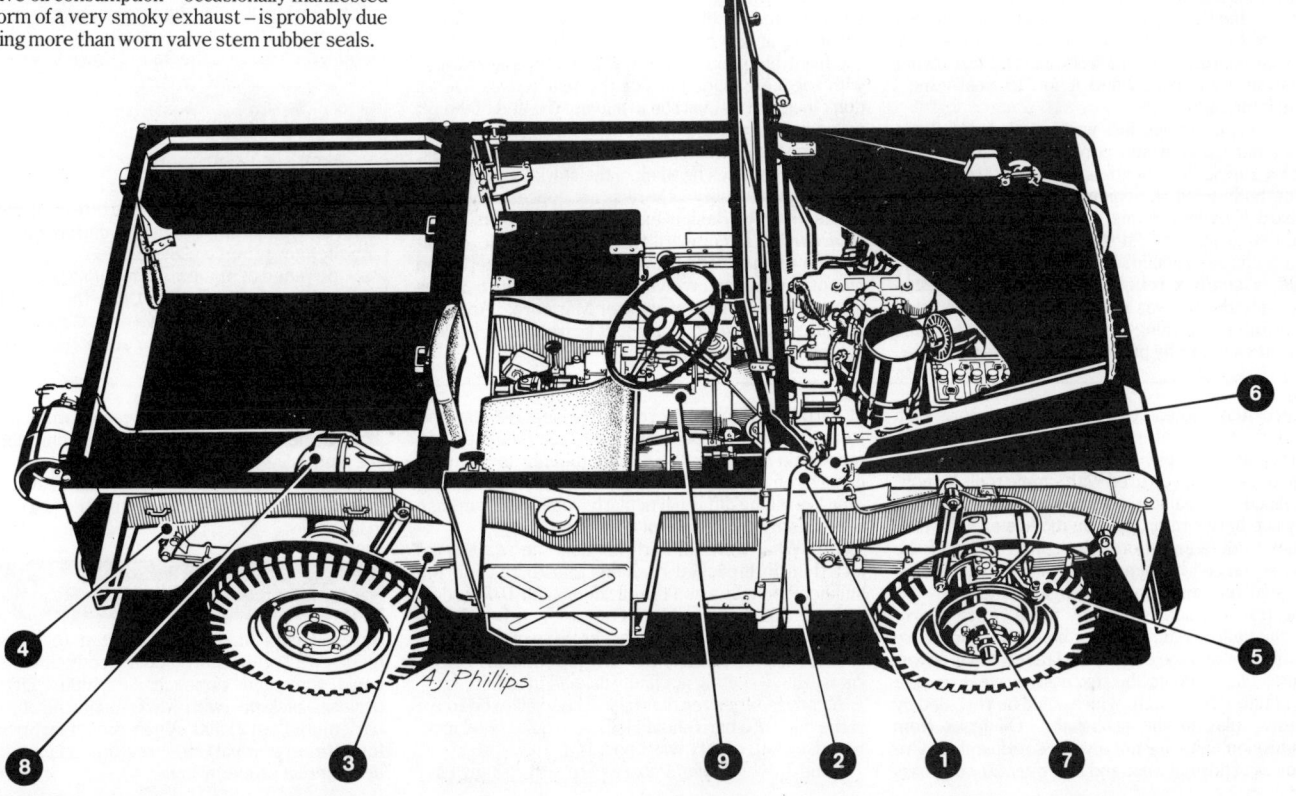

A.J. Phillips

OWNER'S VIEW

Land-Rovers aren't just for country use – London ad-man Julian Moseley reckons his Series 1 takes some beating as daily city transport

Julian Moseley's office is predominantly grey – carpet, furniture, gas fire – but it's littered with old car memorabilia; a hub-cap from a Mercedes, an enamel Cadillac sign, piles of Panhard snapshots, brochures for Norton motorbikes. This is the nerve centre he shares with his partner, Panhard Dyna man Tony Mackertich, and from where their creative advertising concepts emerge. Moseley leans heavily on the table. A few hundred yards away, in the bowels of the Soho NCP car park, you can usually find his immaculate Series I Land-Rover.

But not at the moment; its back axle has died after 40 years of transmission pounding, and Moseley will get a taxi home tonight.

"I drive it every day," he says, cigarette in mouth. "I've driven it around town for two years and I'm never sick of it – it's always exciting."

"I've driven it around town for two years and I'm never sick of it – it's always exciting..."

His other car, a '69 Mustang convertible, is not ideal for London. So he had a Renault 4 as a runabout, and when the floor fell out of it he took the opportunity to buy a Series I Land-Rover, inspired by Dinky Toys' version of the fifties and a taxi trip in one on honeymoon in the Grenadines: "It was the guy's whole livelihood, his workhorse, everything.

"So I asked Geoff at USA GB, who looks after my Mustang, and he suggested a company called Abbey Motors in North London. There was one for sale on its notice board, a running vehicle. The last owner and his brother had owned it for 15 years and I bought it for £400."

However, there was lots wrong with it, like 8in of play in the steering and zero braking ability, and Moseley anticipated spending about £2000 on it.

"The body is all aluminium and the chassis is an Isambard Kingdom Brunel dream, so I thought I'd spend two grand on it; at the end of the day I spent six on a chassis rebuild, a body rebuild, an engine rebuild, a gearbox rebuild, new tyres... in retrospect, only the tilt was sound! The main thing that goes in the damn things is the bulkhead, and mine was pretty sound – by pure luck, not judgement."

Moseley doesn't use it off the road, however. This, apparently, requires driving skills that are the opposite of road driving – "you have to be in reverse when you should be in fourth, braking when you should be accelerating" – skills in which fellow members in the Land-Rover Series I Club are versed.

"They're maniacs, they're obsessed, they'd slobber to see an early 109in estate car with tropical roof. But... the newsletter is exemplary. They have them for completely different reasons; they are fundamentally farm vehicles, and their owners live in rural communities – it probably has great prestige to drive a Series I in Devon."

What is the Land-Rover like to use every day?

"It's nippy and economical but it can be bloody cold. It's fully waterproof, however, as I've treated the tilt with a spray from a camping shop that knits the canvas together. I've driven it with the roof and door tops off, the screen sticks removed and the screen folded but I need to wear glasses because my eyes stream... and there's no rear-view mirror. That's flat on the bonnet, staring at the sky."

He has fitted a brake servo which improves stopping immensely, and a set of winkers and extra sidelights at the front, on special brackets fixed to the front bumper. Also fitted is a concealed stereo with speakers hidden beside the seat squabs. Doors don't lock and a small slip in leaving the end of the tilt rolled up resulted in a stolen starting handle. As someone concerned with images and appearances, though, what does he think of the stark Land-Rover?

"It's pure utility, there's no concession on it whatsoever to design excess, a 2CV looks like a *boulevardier* in comparison. Apart from the curvature on the dash and transmission cover, there's no hint of any human involvement." The flame from his lighter kisses the tip of another Marlboro; "And," he adds through a smoke haze, "it scares the hell out of drivers of black cabs."

The clutch shouldn't show anything more than normal wear, even after long and hard use. Though by virtue of its simple construction the Land-Rover can quite easily be repaired with simple tools by non-specialised mechanics, there was one annoying difference between production models and the pre-production demonstrators; the construction of the transmission bell-housing meant that clutch replacement involved removing the engine entirely.

The transmission is unlikely to be overly worn, since the gearbox and transfer box in all models are well able to stand the engine's relatively small power output. There could be excessive wear in the differentials (8), though, which could be revealed by excessive play in the propshafts. Oil leaks from transmission units are not always serious; if they're the result of normal wear and tear over 20-odd years

of hard use there's no reason to suggest the unit won't last another 20 years if it's treated gently and topped up regularly. Look out, though, for signs of damage on gearbox casing (9) or differential housing – the low-slung diff is particularly vulnerable – in case the leaks are the results of an impact.

Trim: What trim? In early models the seats were just thin fibre-packed cushions laid to rest on the bulkhead box. It wasn't until the 86 and 107 models arrived that foam rubber comfort reached Land-Rover drivers' behinds and backs. So simple was the seating that it's unlikely that any unrestored Series I car will have seating not badly in need of repair.

A canvas tilt cover that hasn't been allowed to rot in the back of a barn should still be in quite good nick, because cloth quality was good.

Land-Rover Series I spares are surprisingly easy

CLUBS

The **Land-Rover Series One Club** caters exclusively for the vehicles featured in this *Profile*. It publishes up to six newsletters a year, and helps with technical queries and parts finding. There's a £1.50 joining fee, and the sub is £7.50 a year (UK), £9 (Europe), £12.50 (rest of the world). For details of social events and membership pack, contact the Secretary, David Bowyer, at East Foldhay, Zeal Monachorum, Crediton, Devon EX17 6DH (tel: 036 33 666, office hours)

The **Land-Rover Register (1947-1951)** covers only the earliest vehicles. It also has a bi-monthly newsletter; annual sub is £8. Contact the Membership Secretary, Sally Cooknell, at High House, Ladbroke, nr Leamington Spa, Warwickshire CV33 0BT (tel: 092 681 2101). For details of other Land-Rover clubs, contact the **Association of Rover Clubs Ltd,** which is based in Rochdale, Lancashire (tel: 0706 30200)

SPECIALISTS

P.A. Blanchard & Co, Foggathorpe, Selby, Yorkshire YO8 7PR (tel: 0757 288613)
Aylmer Motor Works, Wood Green, London N22 (tel: 01-889 9401)
Kingsdown Unit 1, New Road Business Estate, Ditton, Maidstone, Kent ME20 6AF (tel: 0732 844572)
John Craddock Ltd, 70-76 North St, Bridgtown, Cannock, Staffordshire WS11 3AZ (tel: 05435 77207/5408)
Dunsfold Land-Rover Ltd, Alford Road, Dunsfold, Surrey GU8 4NP (tel: 048649 567)

BOOKS

The Land-Rover 1948-1988 by James Taylor (£10.95). Handy and well-illustrated MRP Collector's Guide. Essential reading.
The Range Rover/Land-Rover by Graham Robson (£9.95). Strong on company background; less well illustrated than Taylor's book, but features the Range Rover as well.

ACKNOWLEDGEMENTS

Grateful thanks to David Bowyer and Peter Adams of the Overlander 4 × 4 Centre, used as our photo location. The centre offers tuition in off-road technique over a non-damaging but challenging course (£85 for 1½ days), and sells a wide range of equipment for off-roading; address as the Series One Club. Thanks also to S1 owners Andrew Stephens and Gary Crisp – as well as to David Bowyer, for use of his Station Wagon.

to come by, and they aren't necessarily expensive Though new parts are virtually unobtainable ther are specialist dealers handling used and recond tioned items. If a rare item *doesn't* crop up one of th specialists may be able to have the part special made.

PRICES

Series I cars haven't yet started to attract 'sill money'. A well-restored 80in woody Station Wago could command as much as £4000, but a mor ordinary pick-up, with MoT, ought not to set yo back more than £1500. A non-runner on bricks, ide for a three-year part-time restoration project, cou be yours for around £250.